INTERNATIONAL
Standards Worldwide

SELECTED TECHNICAL PAPERS
STP1586

Editors: Robert E. Chapman, Michael N. Goodkind

Ideas to Impact: How Building Economic Standards Keep You on Track

ASTM Stock #STP1586

ASTM International, 100 Barr Harbor Drive, PO Box C700, West Conshohocken, PA 19438-2959.
Printed in the U.S.A.

Library of Congress Cataloging-in-Publication Data

Symposium on Building Economics (2014 : New Orleans, La.)
Ideas to impact : how building economic standards keep you on track / editors, Robert E. Chapman, Michael N. Goodkind.
pages cm
"ASTM Stock #STP1586."
Proceedings of the 2014 Symposium on Building Economics held in New Orleans in October.
Includes bibliographical references.
ISBN 978-0-8031-7617-1
1. Building--Cost control--Congresses. 2. Buildings--Economic aspects--Congresses. 3. Sustainable buildings--Standards--Congresses. 4. Facility management--Cost effectiveness--Congresses. I. Chapman, Robert E. II. Goodkind, Michael N. III. ASTM International. IV. Title.
TH435.S945 2014
338.4'769--dc23 2014033296

Peer Review Policy

Each paper published in this volume was evaluated by two peer reviewers and at least one editor. The authors addressed all of the reviewers' comments to the satisfaction of both the technical editor(s) and the ASTM International Committee on Publications.

The quality of the papers in this publication reflects not only the obvious efforts of the authors and the technical editor(s), but also the work of the peer reviewers. In keeping with long-standing publication practices, ASTM International maintains the anonymity of the peer reviewers. The ASTM International Committee on Publications acknowledges with appreciation their dedication and contribution of time and effort on behalf of ASTM International.

Citation of Papers

When citing papers from this publication, the appropriate citation includes the paper authors, "paper title", STP title, STP number, book editor(s), page range, Paper doi, ASTM International, West Conshohocken, PA, year listed in the footnote of the paper. A citation is provided on page one of each paper.

Printed in Bay Shore, NY
September, 2014

Foreword

This Compilation of *Selected Technical Papers*, STP1586, *Ideas to Impact: How Building Economic Standards Keep You on Track*, contains peer-reviewed papers that were presented at a symposium held October 5, 2014 in New Orleans, LA, USA. The symposium was sponsored by ASTM International Committee E06 on Performance of Buildings and Subcommittee E06.81 on Building Economics.

The Symposium Chairpersons are Robert E. Chapman, NIST, Gaithersburg, MD, USA, Michael N. Goodkind, Alfred Benesch & Co., Chicago, IL, USA, and Muthiah Kasi, Alfred Benesch & Co., Chicago, IL, USA. The STP Editors are Robert E. Chapman and Michael N. Goodkind.

Contents

Overview

This ASTM Volume, *STP 1586*, serves as the proceedings for the *2014 Symposium on Building Economics*. The papers in *STP 1586* are representative case studies that demonstrate how building economic standards keep your project on track. Sponsored by ASTM Subcommittee E06.81, the Symposium was held at the Sheraton New Orleans, New Orleans, LA, in conjunction with the October ASTM E06 Committee meetings.

ASTM's suite of E06.81 standards provides a comprehensive framework for measuring and managing the economic performance of buildings, building systems and components, and other constructed facilities over their entire life-cycle. The Symposium demonstrated how the various E06.81 standards fit together and reinforce each other. The E06.81 standards framework is organized around four topic areas: (1) cost data presentation and analysis; (2) value engineering; (3) risk management; and (4) economic evaluation.

The first paper introduces the E06.81 Subcommittee standards; it shows how they can assist key project activities and measure their associated economic impacts. The last paper focuses on how to make the standards more useable and more valuable to key stakeholders over the life of a construction-related project. The rest of the papers focus on various life cycle phases of construction projects in the construction industry's key sectors.

There are two papers that focus on Value Methodology (VA -Value Analysis /VE - Value Engineering). The case studies presented demonstrate the favorable impact that the use of VA/VE has on the performance and cost of the projects highlighted.

A key area of interest for the E06.81 Subcommittee is the elemental cost concept and UNIFORMAT II. The E06.81 Subcommittee's goal is to deliver UNIFORMAT II standards that are relevant, useful, and highly valuable to the construction industry. These classifications provide the foundation for many of E06.81's economic efforts. VA/VE revolves around the formulation of the UNIFORMAT II cost framework. The paper on UNIFORMAT II lays out the philosophy and logic of this family of classifications. Another interesting paper related to the elemental cost concept is the European perspective on elemental cost.

There are four papers that focus on key building-related topics. The first paper explains the Job Productivity Measurement practice, showing how it improves on-site productivity performance through better tracking of the time, cost, and quality of construction. The second paper describes the economic impact of NIST's research and development activities that resulted in improved service life prediction for seams in low-slope EPDM roofing. The third paper estimates the economic impact of the adoption of more stringent state energy codes for new commercial buildings. Estimated savings are based on ASTM standards for calculating life-cycle costs and related measures. The fourth paper evaluates the economic performance of residential sprinklers using the ASTM standard practices for calculating benefit-to-cost ratios and net benefits.

Overall, the 10 papers present a comprehensive summary of the ASTM E06.81 Building Economic standards, showing how they can measure and positively impact the economic performance of construction-related projects. The topics covered highlight the benefits of using the E06.81 standards at the planning, design, construction, and operation phases.

Linking ASTM standards with economic and cost risk factors is helpful in addressing today's economic and global challenges facing the construction industry. The papers suggest ways to keep building-related projects on track. The Symposium provides an opportunity for all ASTM technical committees to learn more about the E06.81 standards, their benefits, and how to utilize them.

This Symposium reflects the dedication and accumulated knowledge of many long-standing ASTM E06.81 members. On behalf of E06.81 Subcommittee, we express our thanks to the sponsors, authors, and ASTM staff who have helped to make this Symposium successful.

Robert E. Chapman
Michael N. Goodkind
Muthiah Kasi

STP 1586, 2014 / available online at www.astm.org / doi: 10.1520/STP158620140029

M. Kasi[1]

Benefits of ASTM Economic Standards

Reference

Kasi, M., "Benefits of ASTM Economic Standards," *Ideas to Impact: How Building Economic Standards Keep You on Track*, STP 1586, Robert E. Chapman and Michael N. Goodkind, Eds., pp. 1–22, doi:10.1520/STP158620140029, ASTM International, West Conshohocken, PA 2014.[2]

ABSTRACT

The ASTM E06.81 Building Economics Subcommittee developed 26 standards that can assist in key project areas. Almost every action throughout the life of a project has cost and risk consequences. This paper provides an overview of how ASTM standards are used to promote more cost-effective decisions for the design, construction, operation and disposal of constructed facilities. There are 10 standards focused on compiling, analyzing, reporting, and summarizing cost data; three standard practices for measuring the value of construction related systems and attributes; five standards for measuring and managing risk; and seven standards for measuring economic performance and reporting the results of an economic evaluation. Terminology is an important part of all E06.81 standards. A separate terminology standard provides definitions of economic techniques and terms used in the 25 other E06.81 standards. Five case studies are presented in this paper. Case Studies 1 and 2 focus on risk management of low-rise and high-rise buildings. Case Study 3 is an example of value analysis of high rise building columns. Case Study 4 describes how to manage cost and mitigate risks for bridge projects and Case Study 5 is part of economic evaluation of a long-span bridge from concept to construction and beyond.

Keywords

buildings, bridges, columns, economic performance, risk mitigation, value engineering, function analysis, elemental cost

---

Manuscript received March 25, 2014; accepted for publication June 10, 2014; published online August 5, 2014.

[1]ASTM E6.81 Subcommittee Chairman; Chairman Emeritus, Alfred Benesch & Company, 205 North Michigan Avenue, Suite 2400, Chicago, IL 60601, e-mail: mkasi@benesch.com

[2]ASTM Symposium on *Ideas to Impact: How Building Economic Standards Keep You on Track* on October 5, 2014 in New Orleans, LA.

Introduction

ASTM Subcommittee E06.81 on Building Economics develops Standards that can help all areas and life-cycle phases of a project [1]. Almost every action in a project has cost consequences throughout the life of the project. This paper provides an overview of how ASTM standards are being used to promote more cost-effective decisions for the design, construction, operation, and disposal of constructed facilities.

ASTM Economics Standards are divided into four sections (1) cost data presentation and analysis, (2) value analysis, (3) risk management, and (4) economic evaluation. There are 10 standards focused on compiling, analyzing, reporting, and summarizing cost data; three standard practices for measuring the value of construction related systems and attributes; five standards for measuring and managing risk; seven standards for measuring economic performance and for reporting the results of an economic evaluation. The compilation [1] also includes a terminology standard.

Five case studies are presented in this paper. Case Studies 1 and 2 focus on risk management. Case Study 3 is an example of value analysis for columns of a high-rise building. Case Study 4 describes how to manage cost and mitigate risks for bridge projects and Case Study 5 is part of economic evaluation. These case studies are based on the ASTM Standards listed in **Table 1**.

TABLE 1 ASTM Standards used in this paper.

ASTM Standard	Description of Standard
E2506	Standard Guide for Developing a Cost-Effective Risk Mitigation Plan for New and Existing Constructed Facilities
E2013	Standard Practice for Constructing FAST Diagrams and Performing Function Analysis During Value Analysis Study
E1699	Standard Practice for Performing Value Engineering (VE)/Value Analysis (VA) for Projects, Products and Process
E1804	Standard Practice for Performing and Reporting Cost Analysis During the Design Phase of a Project
E2103	Standard Classification for Bridge Elements—UNIFORMAT II
E1369	Standard Guide for Selecting Techniques for Treating Uncertainty and Risk in the Economic Evaluation of Buildings and Building Systems
E917	Standard Practice for Measuring Life-Cycle Costs of Buildings and Building Systems

A "project" to deliver a facility comes into being when the need for a new, expanded, or modernized constructed facility has been identified. The project delivery process begins when the client identifies the need and concludes when the constructed facility is turned over to the client. The project delivery process, which may last anywhere from a few months to a few years, is not the end of the life-cycle of a constructed facility. The life-cycle of a typical constructed facility often spans many decades and is largely concerned with the operation and maintenance activities required to meet the client's need. Therefore, decisions made during the project delivery process may have significant financial and operational impacts over the course of a constructed facility's life cycle. These impacts are measured via cash flows, both negative due to increased operational expenses and positive due to increased sales of goods and services.

The time-value of money concept is used in order to deal with the different timing of cash flows associated with the constructed facility. Basically, this is done through the use of a discount rate, which equates or discounts dollars occurring in different years to a common time, referred to as the base year. Because the life cycle of a constructed facility may span many decades, it is useful to define the various phases in the life cycle. Commonly used phases in the life cycle of a constructed facility are: (1) planning, (2) programming, (3) design, (4) construction, (5) operations, and (6) disposal.

Protecting constructed facilities from extreme events—fires, floods, earthquakes, and other natural and man-made hazards—is a constant challenge for facility owners and managers. Choosing among alternative protection strategies is complicated by the fact that such strategies frequently have significant up-front investment costs, result in operations and maintenance costs that are spread over many years, and impact key stakeholders in different ways. A methodology is needed to assure that all relevant costs are captured and analyzed via well-defined metrics.

To address this need, ASTM E2506-11 [2], presents a three-step protocol that establishes a methodology for dealing with extreme events. The three-step protocol has the following essential components: (1) risk assessment, (2) identification of potential mitigation strategies, and (3) economic evaluation. Risk assessment is used to identify the risks confronting a facility. It includes development of possible damage scenarios, probability assessments for these scenarios, and identification of the facility's vulnerabilities and critical areas.

Identification of mitigation strategies—engineering alternatives, management practices, and financial mechanisms—provide performance and cost data for the possible combinations of risk mitigation strategies. Combinations of risk mitigation strategies are used to create a candidate set of alternatives for in-depth economic evaluation. The third component, economic evaluation, enables facility owners and managers to evaluate each alternative combination of risk mitigation strategies and the sequence of cash flows associated with their implementation.

Case Study 1: Site Selection

Case Study 1 follows ASTM E2506 [2]. Section 8.1.2.1 states: "*Management strategies can be procedural or technical. Some management strategies relate to security, training, and communications. Others relate to decisions on where to locate the building and who should have access to its systems and subsystems. Some management strategies complement engineering strategies, while others substitute for them.*" The investigation focused on mitigation of risk through management decisions.

In planning a site for a proposed data center, stakeholder requirements and protecting the proposed facility from extreme events were selected as the two most important criteria. Site selection was based on risk management and function analysis (value analysis), ASTM E2013-12 [3]. The investigation includes the following six items:

1. Airplane flight paths
2. Historical tornadoes path
3. Seismic information
4. Security
5. Railroad commuter and freight lines
6. Neighborhood impact

The first three items are investigated for risk management and the last three items are part of assembling needs and desires.

More than 100 possible sites were analyzed for risk potential and 36 sites were narrowed down in phase one screening based on the above five criteria. Selected sites avoided past tornado paths (**Fig. 1**) and also high frequent flight paths. For the data center, assuring security is vital in analyzing various sites and neighborhood

FIG. 1 Historical tornado paths.

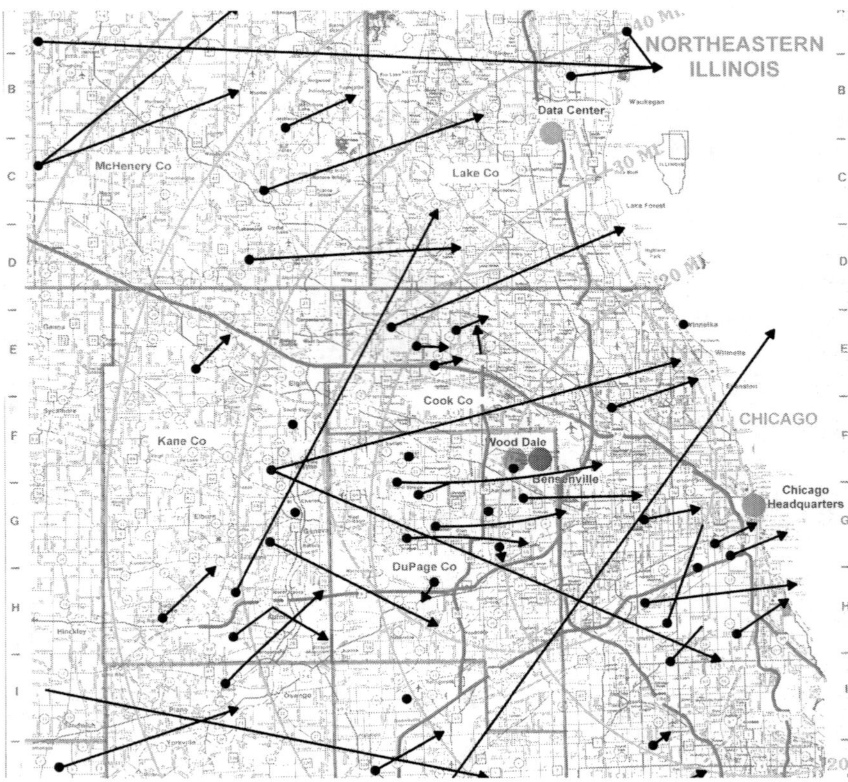

characteristics investigated by visiting some of the sites. The investigation also focused on the terrain, access roads and vacant lots.

The second phase of screening was to select the site that meets all the needs, desires of the stakeholders. **Figure 2** shows the location of the sites that were studied. Potential employee locations are compared with these locations to assess skill and availability. The location of public transportation, local roads and freeways, shopping areas, restaurants, and hotels were noted in the evaluation of the sites. Using the customer based function classification, the final site was selected.

FIG. 2 Site selection map.

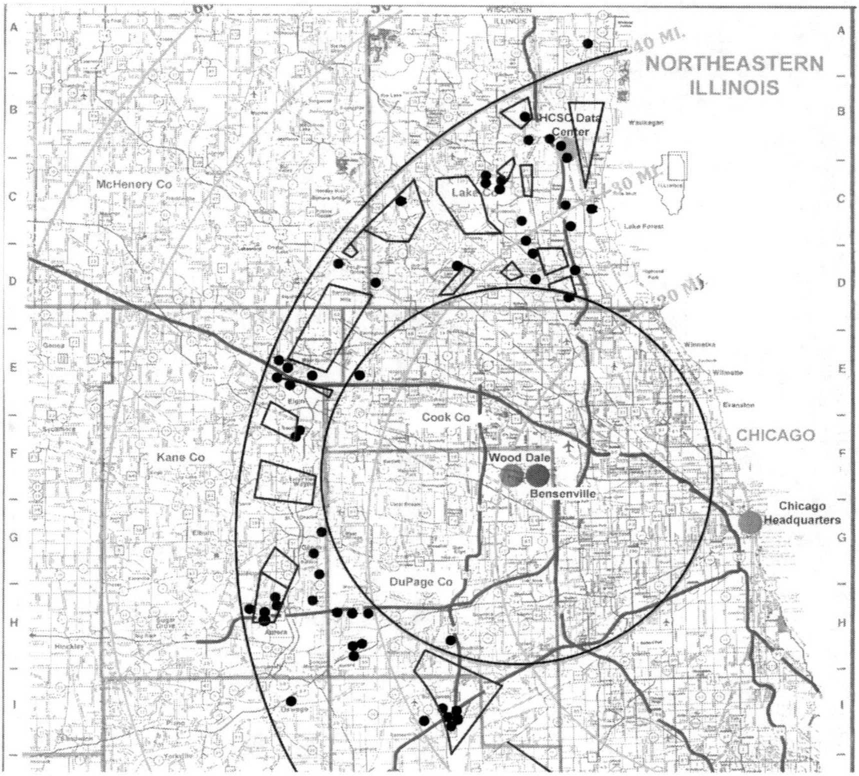

The selected 36 sites were analyzed for technical risks. This included: flooding avoidance, security issues in terms of access, potential delay of firefighting, and other emergency responses. The narrowed selected sites were then analyzed for financial risks which included: potential property tax increase, cost of facility expansion, and lack of qualified employees in the vicinity of the potential site. The final selection fulfilled the needs, desires, and constraints within the limit of the three risks.

Case Study 2: Mitigating Risks (High-Rise Building)

This study is another example of mitigating engineering, management and financial risks as defined in ASTM E2506 [2].

In a 60-story high-rise building, the owner planned to mitigate potential risks. Economic, engineering and management risks were explored. A 12 day VE study, ASTM E1699-13 [4], included the location of the headquarters, number of floors, width of sidewalk, and type of structure. Based on the return on investment and the usage of the building, the decision was made to phase the construction with 30 stories immediately and a proposed vertical expansion in the future. Foundation and columns were designed to accommodate future expansion. When the need for higher occupancy occurs, the owner can initiate a vertical expansion. This mitigated the potential economic risk.

The second point of the study addressed the risks due to potential terrorist attack. The security consultants developed procedures to prevent access to anyone trying to damage the building. This includes permitting vulnerable area access and parking in vulnerable areas to security cleared personnel. These are part of the management decisions to mitigate any risk (see **Fig. 3**). People precautions are part of the management risk mitigations. Structural precautions are part of the engineering mitigation strategy.

FIG. 3 Mitigation of management and engineering risks.

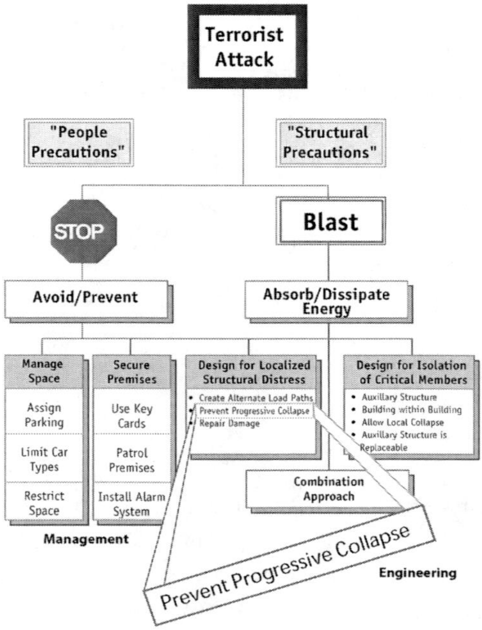

The study further addressed the risk mitigation due to engineering decisions and various scenarios of potential blasts in the proximity of the building. **Figure 4** demonstrates one such potential blast. This is based on the predetermined location of the curb or barrier at the road level. **Figure 5** indicates the expected pressure at the face of the building. **Figure 6** demonstrates that the expected failure of certain

FIG. 4 Building proximity of column to bomb location.

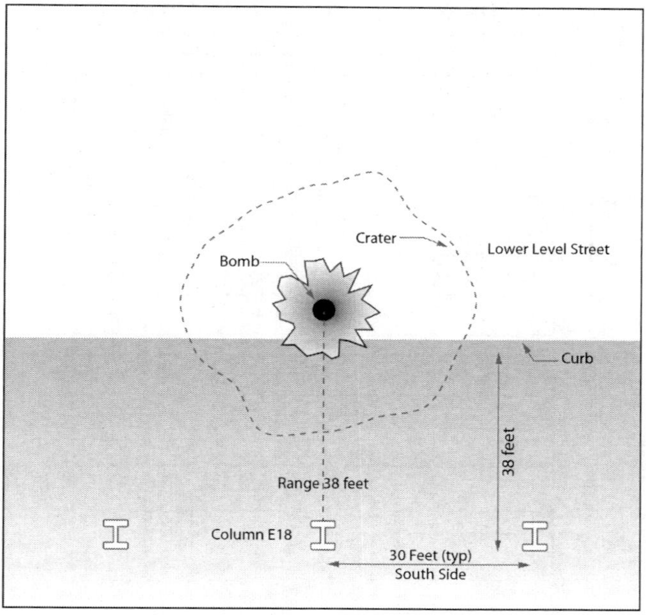

FIG. 5 Expected pressure at the face of the building.

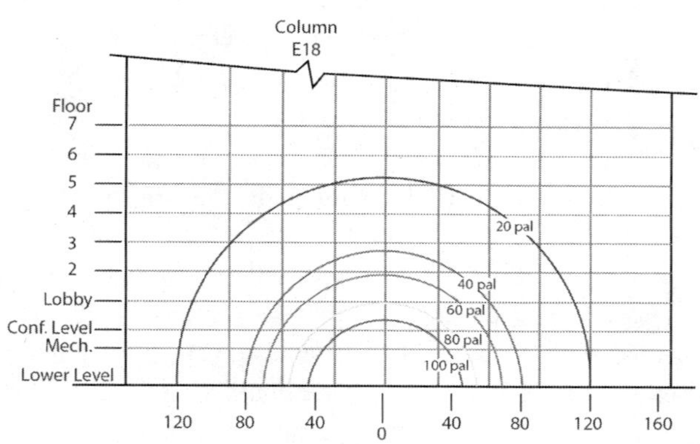

FIG. 6 Expected failure of structural elements.

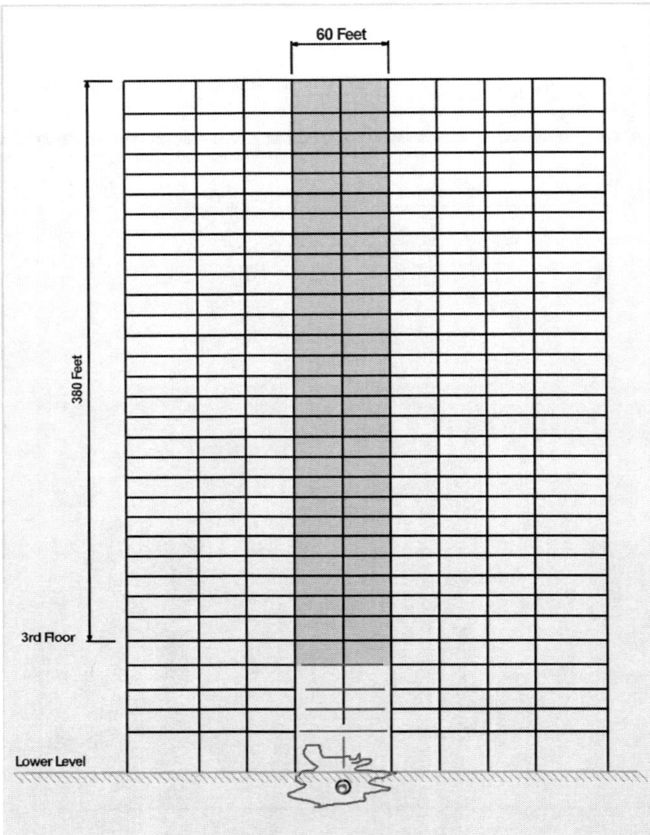

structural elements. The failure of these elements initiates a progressive failure of the building. For different scenarios, strengthening of adjacent structural members would prevent or delay the failure of the partial or full collapse. The decision was made to implement it if it was economically feasible.

Similarly, the design team addressed other potential risks. They developed a management, engineering, and financial risk mitigation strategy to upgrade the reliability of the building.

For a scenario similar to the Oklahoma blast, this building had many positive features. This increased the effectiveness of the building against such a disaster. A bomb loaded truck located along the street would not cause this building to collapse. However, no matter what the size of the bomb, there will always be some localized damage such as: façade destruction, damage to interior slabs and ceilings, and some secondary framing members. This is based on the conceptual theory of relative strength of building elements of the two buildings by comparison. A rigorous analysis is needed to confirm the findings.

Case Study 3: Value Engineering of High-Rise Building Columns

Economics deals with the efficient allocation of resources. In the context of constructed facilities, these resources include requirements, plans, labor, materials, equipment, information, and physical components, all of which have cost associated with them. Thus, economic standards help decision makers choose more cost-effective combinations of resources.

The cost effective combination of resources can be achieved by conducting a value analysis study as prescribed in ASTM E1699 [4]. This case study focuses on what is described in section 5.4.2 of determining the relevance of a requirement.

5.4.2 Perform VE/VA during preliminary design to analyze the relevance of each requirement and the specifications derived from it. Critically examine the cost consequences of requirements and specifications to determine whether the resultant cost is comparable to the worth gained. Further analyze high-cost, low performance or high risk functions and the identification of alternative ways of improving value.

A value engineering study was conducted for a high-rise building. The proposed building was a 35-story office building. One of the elements that was studied was the building columns. There were 1400 columns in the building (40 columns per floor). The study addressed the functions and their cost for the columns.

Columns have the following functions:

- Receive load
- Resist load
- Resist compression
- Resist tension
- Resist buckling
- Transfer load
- Facilitates partition

ASTM E2013 [3] defines value and mismatches as follows: "*Compare function cost and function preference.* **Table 2** shows four possible combinations of cost/

TABLE 2 Illustration of cost/preference combinations.

Type of Combination	Function Cost	Function Preference	Function Value
A	High	Low	Mismatch (Mandatory Value Analysis) (VA)
B	High	High	Candidate for VA
C	Low	Low	
D	Low	High	Match

preference. If the cost of a function is high and the stakeholder thinks its importance is low, the result is a mismatch (Type A in **Table 2**). On the other hand, if the cost of a function is low and the stakeholder rates its importance as high, a high value is achieved and the stakeholder has a match (Type D in **Table 2**). These are the two extremes of the cost/preference measurement."

The function of the tie in square column is to resist buckling. A circular column has inherent stiffness to resist buckling without any ties. Both resist buckling; however, the cost to satisfy this function is less for circular columns. In this case, circular columns have value while square columns lack value, which is a mismatch (see **Fig. 7**).

FIG. 7 Function–resist buckling.

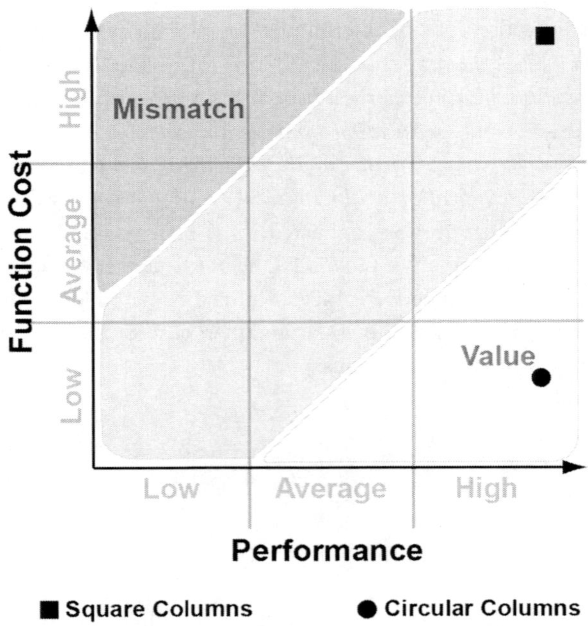

Traditionally, square columns are preferred in high-rise buildings. The reason behind this practice is that the construction of partition wall around rectangular or square columns is much simpler. It is one of the honest wrong beliefs. Contractors prefer to enclose either square or circular columns with a furring (drywall enclosure). This gives the contractor the flexibility of constructing the column with a larger tolerance for plumbing.

Various combinations of shape, concrete strength, and column size were designed and cost was computed. It revealed circular columns were less expensive to construct. A higher cost of the square column is the attributed to formwork and reinforcing. For any particular column, the formwork cost was $259.00 for a square column and $94.00 for a circular column. **Table 3** shows the properties and cost detail of a square column and **Fig. 8** shows the "as designed" square column reinforcing detail. **Table 4** shows the properties and cost detail of a circular column and **Fig. 9** shows the proposed circular column reinforcing detail. The cost of the formwork and reinforcing are reduced considerably.

TABLE 3 Properties and cost detail of square column.

	Square Column		
Size	914 by 914 mm (36 by 36 in.)		
Strength	41.37 MPa (6000 psi)		
Cost per column per floor	Concrete	$270.00	31 %
	Formwork	$259.00	30 %
	Reinforcing	$337.00	39 %
	Total	**$866.00**	

FIG. 8 Reinforcing detail of square column.

TABLE 4 Properties and cost detail of a circular column.

	Circular Column		
Size	914 mm (36 in.) diameter		
Strength	62.05 MPa (9000 psi)		
Cost per column per floor	Concrete	$295.00	31 %
	Formwork	$94.00	30 %
	Reinforcing	$169.00	39 %
	Total	**$558.00**	

FIG. 9 Reinforcing detail of circular column.

The cost of column per floor was reduced from $866.00 to $558.00. Similarly every column was analyzed for various sizes and strengths as a circular column. In computing various concrete strengths, shapes, and sizes for a given load (**Fig. 10**), it is evident that more choices are available. Resisting compression can be achieved by a combination of concrete strength, reinforcement, and size of the columns. Higher-strength columns require less size and will result in lower cost. As column size increases the effect of higher strength diminishes, resulting in increased cost. For the same strength of concrete and size of the column, circular column costs less than square ones due to simplified formwork. With prefabricated formwork for circular columns, it requires less labor to form. These factors reduced the cost of the circular columns.

FIG. 10 Comparison of various concrete strengths, shapes and sizes for a given load.

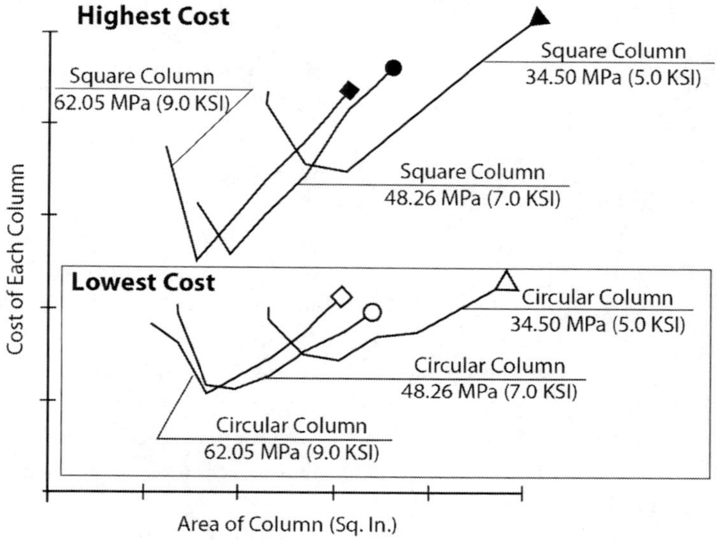

A value analysis process [4] can lead a team into looking at all possibilities.

The 35-story building with 40 columns per floor resulted in $365 000 savings to the owner. This led to a change in the column design during construction. **Figure 11** shows the change of column from square to circular after the VE study.

FIG. 11 Building with the change in column shape.

As demonstrated here, a value methodology [4] can be adapted at any phase of the project. In the beginning, it is called value planning; during construction, as in this case, it is called value engineering change proposal (VECP).

Case Study 4: Managing Cost and Mitigating Risks (Bridge)

This study followed the ASTM E1804-12 [5] for managing cost through the six phases as described in this paper.

A proposed innovative design of the Gateway Arch Bridge at the schematic phase raised concerns by the Federal and State of Michigan governments due to a significant number of unknowns and unfamiliar conditions. The designers conducted a number of function analysis studies using ASTM E1699 [4] and ASTM E2013 [3] and identified possible impact elements and functions. These included the concrete deck type, hanger type, foundation type, and the shape of the arch. The Gateway Arch Bridge is a dual single-span, modified tied-arch carrying six lanes of Interstate 94 (I-94) traffic (three eastbound and three westbound) over Telegraph Road in Taylor, MI.

Due to the uniqueness of this structure, using cost data from past projects did not guarantee effective cost management. Using ASTM E1804 [5] and ASTM E2103-11 [6], cost variation was tracked with proper documentation (see **Table 5**). ASTM E1804 was followed in all phases of the project.

TABLE 5 Cost at project phases.

Phase/Cost Item	Cost[a]	Reasons/Comments
Base cost	$6 870 000	
Non-specific contingency	$1 200 000	Maintain $1.2 M as contingency
***Program Phase* Subtotal**	$8 070 000	
Increase span length	$70 000	Better sight distance
Escalation of steel price	$120 000	Uncertain/unit prices escalation
Change from tied arch to Modified arch	$200 000	Introduce grade beam ties at road level to increase safety
Provision for post tensioning	$20 000	Add provision for future maintenance
***Schematic Phase* Subtotal**	$7 280 000	
Change from post tensioned to Conventional deck	$154 000	Change post-tensioned deck to conventional concrete deck. Not comfortable with post-tensioning
Change rods to strands	$100 000	Easier to transport
Add second set of hangers	$90 000	Add multiple hangers to increase redundancy

TABLE 5 *(Continued)*

Phase/Cost Item	Cost[a]	Reasons/Comments
Revise girder shape from box to Plate in the middle	-$250 000	
Optimize floor	-$79 000	Reduce longitudinal beam
Overlay, barriers and Landscaping	$125 000	
Pressurize ribs	$180 000	Minimize maintenance
***Design Development Phase* Subtotal**	$7 600 000	
Optimize/balance arch rib Geometry	-$420 000	Efficient use of material and method by design
***Construction Documents Phase* Total**	**$7 180 000**	

[a]Cost shown here is for one bridge. The project has twin bridges.

The design phase is often subdivided into three sub-phases: schematic design, design development, and construction documents. Schematic design establishes the general scope, conceptual design, and the scale relationships among the parts of the project. The primary goal is to clearly define a feasible concept within the allocated budget in a form that the client understands and approves before proceeding to design development.

During the design phase, a series of value engineering studies were completed focusing on the performance of functions. It began with the elemental cost of the project. Elemental cost provides a better perspective of how cost is allocated.

In the program phase, the elements are not completely defined. For example, the clear distance between abutments is not known until roadway geometry is completed. The base cost was estimated to be $\$6.87 \times 10^6$; the program phase cost estimate included a non-specific contingency of $\$1.20 \times 10^6$. Thus, the total estimated cost for the program phase was $\$8.07 \times 10^6$.

Figure 12 is a graphical representation of cost distribution [6]. There are two high cost elements: the arch rib and the transverse girders. There is one low cost

FIG. 12 Cost distribution of group elements and individual elements.

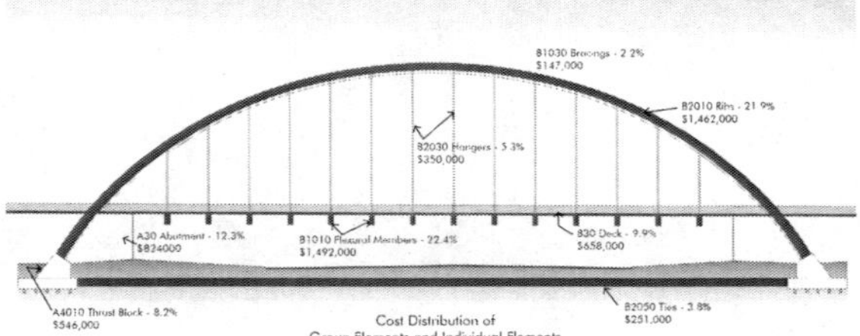

element that is a critical member. Through the value engineering process, the hanger assembly was strengthened and the high cost elements were reduced.

In the schematic phase, the elements are developed. As the elements were developed, the nonspecific contingency was replaced with specific allowances for each cost item. A risk analysis based on ASTM E1369-11 [7] showed the need to address the increasing steel price trend at the time. To better manage this risk, the unit prices of structural steel members for the arch ribs and transverse beams were increased.

There is high level risk in having a tie at higher level on grade separation structures which are vulnerable to damage (see **Fig. 13**). An innovative solution of introducing a tie below the roadway (see **Fig. 14**) was proposed. The steel ties were changed to concrete ties.

FIG. 13 Tie at road level on at-grade separation structure.

FIG. 14 Tie buried under the roadway.

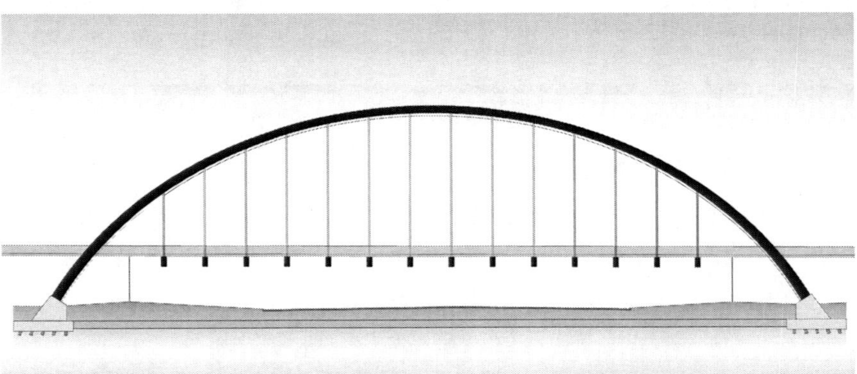

This increased the cost by \$200 000. The foundation tie was assumed to be post-tensioned concrete. To add redundancy, conventional reinforcing to resist the force was added. This increased cost by \$20 000 (management strategy) [2]. The total estimated cost at the end of the schematic phase was \$7.28 $\times$ 10^6.

The hanger assembly went through some changes. First, the type was changed from rods to strands. The cost increase was \$100 000. Because the hanger assembly is a critical member in carrying the load, it was decided to increase its redundancy. Thus, each hanger assembly has a pair of strands, each of which is capable of carrying the total load (see **Fig. 15**). This increased the cost by \$90 000 (engineering strategy) [2].

FIG. 15 Hanger assemblies.

The value engineering process converts challenges into opportunities [8]. The arch ribs are not large enough for inspectors to crawl inside to inspect. For inspection purpose, the box-shaped ribs are pressurized and pressure gages are installed. Inspectors can monitor the pressure for any leakage. If the pressure is maintained, there are no cracks in the welds of the arch. There can be no corrosion inside the welded box shape if no air can bring moisture inside. The inspectors then know that the structure is in good condition. All connections to the rib are complicated due to the pressurization of the ribs. It was a challenge to estimators to compute the cost of these nontraditional connections and curved bracings.

In the construction documents phase, when all policy decisions were made and all details were known, the design team developed a detailed cost estimate compatible to design details and specifications. Changing the arch rib geometry saved \$420 000. The total estimated cost at the end of the construction documents phase was \$7.18 $\times$ 10^6. By using ASTM standards, cost was managed better, contributing to a final engineering estimate of \$7.18 $\times$ 10^6, which is \$890 000 less than the program cost.

Case Study 5: Cost Management from Planning to Disposal

The example project is a proposed bridge over the Mississippi River between East St. Louis, IL and St. Louis, MO. Ten alternative bridge types were considered in the life-cycle cost analysis (LCCA) [9,10], including both cable-stayed and suspension, some that spanned the river (**Fig. 16**: a 2000 ft (609.6 m) span) and others having a pier in the river (**Fig. 17**: a 1500 ft (457.2 m) navigation span). The design service life

FIG. 16 River spanning alternative - no pier in river (Alternative 7).

FIG. 17 Non-river spanning alternative - pier in river (Alternative 6).

for all alternatives was 150 years. The study period for the LCCA was taken as 75 years to account for a complete deck replacement event. The study was based on ASTM E917-05 (2010) [10].

In ASTM E917 [10], the life-cycle cost (LCC) method measures, in present-value or annual-value terms, the sum of all relevant costs associated with owning and operating a constructed facility over a specified period of time. The basic premise of the LCC method is that all costs arising from an investment decision are potentially important to that decision maker, including future as well as present costs. The LCC method compares alternative, mutually exclusive designs or system specifications that satisfy a given functional requirement on the basis of their life-cycle costs to determine, which is the least-cost means (i.e., minimizes life-cycle cost) of satisfying that requirement over a specified study period. With respect to the base case, an alternative is economically preferred if, and only if, it results in lower life-cycle costs.

The initial costs for alternatives vary from $\$287 \times 10^6$ to $\$537 \times 10^6$. Significant differences occur between spanning alternatives and non-river spanning alternatives with respect to routine user costs and vulnerability costs (agency and user). The routine user costs can be as much as two times greater for the non-river spanning alternative than for river spanning alternative.

However, the more significant cost difference is associated with the vulnerability costs. Vulnerability costs result from extraordinary hazards such as seismic events, vessel collisions, or flooding. The river spanning alternatives are vulnerable only to a seismic event and incur similar agency costs. Non-river spanning alternatives are vulnerable to seismic, scour, and vessel collisions, resulting in agency costs three-times more than the river spanning alternatives. The majority of these costs are due to the event, vessel collision. During a major hit the bridge is closed for 24 h, requiring traffic to be detoured onto nearby highways. This results in significant work zone detour user costs. In addition, the barge owner incurs significant costs, which are added with the user costs. Comparing Alternatives 7 and 6, there is large difference in the initial cost and vulnerability costs. However, the LCC places these costs in equal position of the total cost. **Table 6** tabulates these costs for various alternatives.

TABLE 6 Agency and user cost comparison.

	Routine Cost			Vulnerability Cost			
	Agency						
Alternative	Initial Construction	Inspection/ Maintenance	User	Agency	User	Summation of Life Cycle Costs (LCC)	Ranking by Total LCC
ILA2	$379 071 017	$56 575 571	$116 598 227	$179 687	-	$552 425 000	2
ILA3	$385 506 194	$59 734 408	$202 626 990	$179 687	-	$648 049 000	9
IDA2	$373 244 290	$56 973 815	$116 598 227	$179 687	-	$546 996 000	1
ISA3	$379 932 822	$59 687 995	$202 626 990	$179 687	-	$642 427 000	8

TABLE 6 *(Continued)*

	Routine Cost			Vulnerability Cost			
	Agency						
Alternative	Initial Construction	Inspection/ Maintenance	User	Agency	User	Summation of Life Cycle Costs (LCC)	Ranking by Total LCC
8	$536 815 184	$79 041 704	$17 352 821	$178 621	-	$623 388 000	7
2	$484 659 227	$61 997 742	$189 652 267	$553 050	$10 304 310	$747 167 000	10
7	$512 163 221	$80 026 623	$18 627 838	$178 621	-	$610 996 000	5
6	$311 833 135	$60 150 412	$232 928 839	$556 389	$10 304 310	$615 773 000	6
3Y	$286 760 373	$58 778 345	$232 706 857	$553 050	$10 304 310	$589 103 000	4
3D	$315 279 455	$35 899 142	$218 458 499	$554 050	$10 304 310	$580 494 000	3

Inclusion of these routine and vulnerability costs (brought up to present value) reduces the cost difference between cable stayed river spanning alternatives and non-river spanning alternatives.

Another factor that must be considered is the sensitivity of discount rates. The base analysis used a real discount rate of 4 %. To test the sensitivity of this rate, the LCCA was performed using a real discount rate of 2.8 %. **Table 7** illustrates the effects of changing the discount rate.

TABLE 7 Summary of comparative life cycle costs variation in discount rate dollars in millions.

Alternative	Base Discount Rate 4 %	Rank	Base Discount Rate 2.8 %	Rank	Base Discount Rate 7 %	Rank
ILA2	$552.4	2	$728.8	4	$421.4	5
ILA3	$648.1	9	$906.2	8	$449.9	7
IDA2	$546.9	1	$723.4	3	$415.9	4
ISA3	$642.4	8	$900.6	7	$444.3	6
8	$623.4	7	$669.9	2	$571.6	10
2	$742.2	10	$991.8	10	$554.3	8
7	$610.9	5	$659.7	1	$557.5	9
6	$615.7	6	$918.7	9	$385.7	3
3Y	$589.1	4	$890.4	6	$360.4	1
3D	$580.5	3	$853.2	5	$375.9	2

Under the base analysis (DR = 4 %), the cable stays Alternatives 1DA2 and 1LA2 (steel orthotropic main span) have the least costs. A discount rate of 2.8 % changes the overall ranking, resulting in the bridge Alternatives 7 and 8 as the least costs. All four alternatives have a steel orthotropic deck. The event influencing the ranking is cable stay replacement, which has significant agency and user cots. Although this occurs at year 75, a lower discount rate increases the influence of this event to the overall costs. Conversely, increasing the discount rate to 7 % lessens the influence of the later occurring events. In this analysis, Alternatives 3 Y and 3D, the cable stay with the concrete main span are the least costs. Those alternatives with the higher initial costs have the higher life cycle costs. **Figure 18** shows what happens to the life-cycle costs for vessel collisions when the real discount rate varies. For example, increasing the real discount rate from 4 to 8 % will decrease the vulnerability user cost for Alternative 2 from $10 304 310 to $5 735 110.

FIG. 18 Variation in real discount rate for vessel collisions.

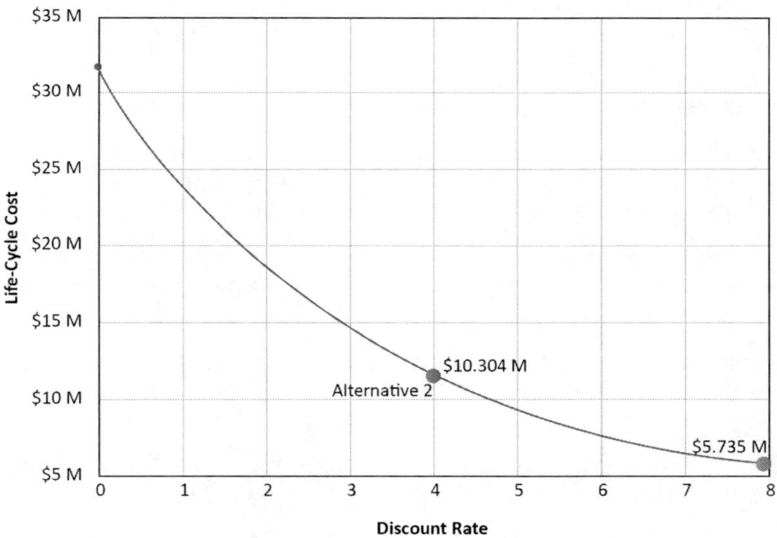

Major bridges are expected to serve the transportation needs for more than 150 years. Over such a long time period they are designed to withstand unexpected events such as seismic activity and still remain functionally adequate and structurally sound. All possible alternatives, through the VE process, should be tested for risks and functions. Through the life cycle cost process they should be measured for cost effectiveness.

Summary

These are some of the applications of the E06.81 Economic standards. It should be noted that all ASTM standards have an impact on costs and benefits. When users apply various ASTM standards such as in sustainability, energy efficiency, etc., economic standards can facilitate the selection process.

Owners and designers can utilize these ASTM standards to create cost effective and safer buildings. Insurance companies can measure the vulnerability of any building with the risk mitigation standards. Contractors can reduce the construction cost using the standards. Transportation officials have consistently used value engineering, risk management, and economic evaluation for asset management and long-term programming. This symposium contains numerous papers that will elaborate on these standards.

References

[1] ASTM International, *ASTM Standards on Building Economics*, 7th ed., ASTM International, West Conshohocken, PA, 2012.

[2] ASTM E2506-11: Guide for Developing a Cost-Effective Risk Mitigation Plan for New and Existing Constructed Facilities, *Annual Book of ASTM Standards*, ASTM International, West Conshohocken, PA, ASTM International, 2013.

[3] ASTM E2013-12: Practice for Constructing FAST Diagrams and Performing Function Analysis During Value Analysis Study, *Annual Book of ASTM Standards*, ASTM International, West Conshohocken, PA, 2012.

[4] ASTM E1699-13: Practice for Performing Value Engineering (VE)/Value Analysis (VA) for Projects, Products and Process, *Annual Book of ASTM Standards*, ASTM International, West Conshohocken, PA, 2013.

[5] ASTM E1804-12: Standard Practice for Performing and Reporting Cost Analysis During the Design Phase, *Annual Book of ASTM Standards*, ASTM International, West Conshohocken, PA, 2012.

[6] ASTM E2103-11: Standard Classification for Bridge Elements—UNIFORMAT II, *Annual Book of ASTM Standards*, ASTM International, West Conshohocken, PA, 2012.

[7] ASTM E1369-11: Standard Guide for Selecting Techniques for Treating Uncertainty and Risk in the Economic Evaluation of Buildings and Building Systems, *Annual Book of ASTM Standards*, ASTM International, West Conshohocken, PA, 2011.

[8] Kasi, M., Function Approach to Transportation Projects—A Guide to Value Engineering, iUniverse, Bloomington, IN, 2009.

[9] National Cooperative Highway Research Program, "Bridge Life-Cycle Cost Analysis," *NCHRP Report* 483, Transportation Research Board, Washington, D.C., 2003.

[10] ASTM E917-05 (2010): Practice for Measuring Life-Cycle Costs of Buildings and Building Systems, *Annual Book of ASTM Standards*, ASTM International, West Conshohocken, PA, 2012.

STP 1586, 2014 / available online at www.astm.org / doi: 10.1520/STP158620140032

Stephen J. Kirk[1]

Enhancing Value for the Guggenheim Abu Dhabi Museum via Function-Inspired Change

Reference

Kirk, Stephen J., "Enhancing Value for the Guggenheim Abu Dhabi Museum via Function-Inspired Change," *Ideas to Impact: How Building Economic Standards Keep You on Track*, STP 1586, Robert E. Chapman and Michael N. Goodkind, Eds., pp. 23–34, doi:10.1520/STP158620140032, ASTM International, West Conshohocken, PA 2014.[2]

ABSTRACT

This paper highlights the principals of value management applied to the planned Guggenheim Museum in Abu Dhabi. The core of success was the use of function analysis, which is the unique element of the value methodology (VM). Both performance and life-cycle cost savings resulted from function inspired change. The techniques of VM as well and the resulting recommendations for value enhancement of the Guggenheim are described. The project is being managed by the Abu Dhabi Tourism Development & Investment Company (TDIC). The lead designer is Gehry Partners (Los Angeles, CA). The schematic VM workshop focused on a review of the site and building design efficiency, architectural and engineering systems effectiveness, sustainability, and life-cycle performance, and risks associated with the schedule and constructability. The VM team identified over 300 ideas for value enhancement.

Keywords

buildings, construction, cost data, economic evaluation, function analysis, FAST, life-cycle costing, Miles Value Foundation, SAVE International, value engineering, value analysis, value management, value methodology

Manuscript received March 31, 2014; accepted for publication June 16, 2014; published online July 14, 2014.
[1]Ph.D., FAIA, CVS-Life, FSAVE, LEED, AP, President, Kirk Value Planners, Kirk Associates, LLC, Goodyear, AZ 85395, e-mail: kirkassociates@aol.com
[2]ASTM Symposium on *Ideas to Impact: How Building Economic Standards Keep You on Track* on October 5, 2014 in New Orleans, LA.

Introduction

According to SAVE International, the internationally recognized authority for value engineering/value analysis (VE/VA), the value methodology (VM) is a systematic process used by a multidisciplinary team to improve the value of a project through the analysis of its functions [1]. In 1993, The Office of Management and Budget (OMB) prepared Presidential Directive A-131 [2] for guidance in application of VE for effectively "reducing costs, increasing productivity, and improving quality." In 1996, Public Law 104-450 was passed by the U.S. Congress, which requires each executive agency to establish and maintain a VE procedure for applying VE to their projects, products, and processes [3].

Lawrence D. Miles, the originator of the VA/VE process, states best value is determined by two considerations: performance and cost [4].

Value is defined as a fair return or equivalent in goods, services, or money for something exchanged. Value is commonly represented by the relationship:

$$\text{Value} \approx \text{Function/Resources}$$

where:

function = performance requirements of the customer and resources measured in materials, labor, price, time, energy, life-cycle cost, etc. required to accomplish that function.

A value methodology focuses on improving value by identifying alternate ways to reliably accomplish a function that meets the performance expectations of the customer [5].

Figure 1 illustrates the various ways value can be added to a project. For example, the arrows in the lower right corner indicate an idea that maintains performance and reduces costs. The arrows in the center indicate an idea that raises

FIG. 1 Value-enhancement options.

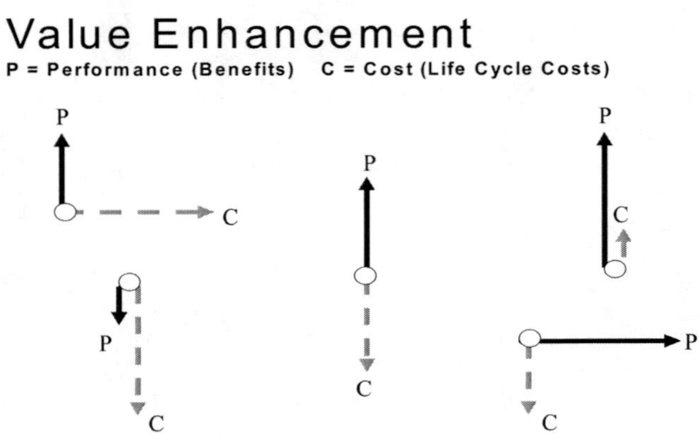

performance and lowers cost. Ideas of all five types are typically identified in a VM workshop [6].

Standard VE/VA Practice

The standard practice of performing VE/VA for projects, products, and processes is defined in ASTM E1699-13 [7]. According to this document, VE/VA is typically applied during the schematic stage of a building construction project or about 15 % design completion. A certified value specialist (CVS) guides a value study. Experience has shown that project studies performed by a person or team with little or no VM leadership will tend to steer in the direction of a superficial review and concentrate on errors made by others. A VM study, on the other hand, focuses on both reducing the total cost of ownership and improving overall performance. Application of the VM methodology and coordination of the activities before and after the study also significantly increase the probability the recommendations will be implemented. The VM job plan workshop methodology [1] consists of the following six steps:

1. information phase,
2. function analysis phase,
3. creative phase,
4. evaluation phase,
5. development phase, and
6. presentation phase.

Function analysis is the foundation of a value methodology and is the key activity that differentiates this body of knowledge from other problem-solving or improvement practices. During the function analysis phase of the job plan, functions are identified that describe the work being performed within the scope of the project under study. These functions are described using two-word, active verb/measurable noun pairings; for example, one function of a museum is to "enhance culture." The team reviews the project's functions to determine those that could be improved. These may be project functions that seem to be performed inefficiently or with more than expected cost. These functions become the focus of the value methodology team in their endeavor to improve the project.

Function analysis systems technique (FAST) diagrams are helpful in organizing functions together in a how–why logic pattern [8]. This helps assure the team that all required functions have been identified. For the museum example, how can the function "enhance culture" be achieved? Answer, by "establishing world-class art galleries."

Life-cycle costing is an important tool during the development phase [9]. It is used to compare the original design with alternatives. Economic criteria such as the discount rate and study period are analyzed. Building costs evaluated in a life-cycle cost analysis include initial capital cost, replacement costs, and annual cost for maintenance, energy, water, etc. Present value (present worth) calculations convert the monies spent at various times to an equivalent cost as of today for comparison of alternatives [10].

Project Description

The Guggenheim Museum Abu Dhabi project consists of the design of a world-class facility for the exhibition of contemporary visual art located in the cultural district of Saadiyat Island in the Emirates of Abu Dhabi, which is part of the United Arab Emirates. The main goal of the project is to bring contemporary visual art from all over the world to Abu Dhabi, in an extraordinary setting at the tip of Saadiyat Island, as part of a series of cultural institutions such as the Louvre Museum, a Performing Arts Center, a Maritime Museum, and the Sheikh Zayed Museum.

The museum will contain exhibition spaces of various sizes and character, for permanent and temporary collections, as well as the support and administrative functions necessary to operate a facility of this size. The building will also contain public amenities such as an auditorium, a restaurant, a café, and a bookstore. The total building is approximately 35 000 square meters (SM).

SCHEMATIC DESIGN

The schematic design is configured to allow for visitors to experience the museum in a unique way—blurring the lines of indoor and outdoor spaces. At the heart of the building is a centralized courtyard, which serves as the primary orientation space for the museum visitor. The courtyard is intended to be cooled through passive and active means providing a comfortable environment for patrons visiting the museum. The exhibition spaces, which are the primary means of circulation are accessed from the central courtyard, are comprised of the three main elements: gallery spaces for a permanent collection, special exhibitions, as well as a series of exhibition spaces creating a center for Arab, Islamic, and Middle-Eastern culture.

The architectural language of the museum is a series of block forms delineating the interior gallery volumes. The cone shapes (see **Fig. 2**), which define the exterior covered spaces around the museum, serve the functions of creating a microclimate environment around the museum as well as providing transitions from indoor/outdoor experiences for visitors moving between temperature- and humidity-controlled gallery spaces and the potentially extreme environment of Abu Dhabi (see **Fig. 2**).

Value Management Study

VALUE MANAGEMENT WORKSHOP OBJECTIVES

The project owner, the Abu Dhabi Tourism Development & Investment Company (TDIC), initiated a VM study at the schematic design phase. Kirk Associates led the study. The following is a summary of the VM objectives:

FIG. 2 Guggenheim Museum Abu Dhabi, UAE, Gehry Partners, LLP.

- improve site and building design efficiency,
- enhance architectural and engineering systems effectiveness,
- further sustainability and life-cycle performance,
- identify project risks and mitigation strategies (constructability/schedule), and
- explore ways to meet the owner's construction budget and schedule.

VM TEAM

The VM team consisted of members from the:

- TDIC, owner,
- Guggenheim Foundation, user,
- Gehry Partners, architect,
- Adamson, executive architect,
- Arup, engineering,
- AECOM, project manager, and
- Kirk Associates, value specialists and facilitators.

More than 50 participated in this 3-day VM workshop.

VALUE MODELS

As part of the information phase of the VM workshop, a number of value models were prepared including cost benchmarks, a cost model, a risk model, a sustainability model, a constructability checklist, and a schedule (time) model. Each model helped the team identify ideas for value enhancement.

Cost Benchmarks

The VM team compared the cost per square meter of the Guggenheim Abu Dhabi museum with other recent international museums. All project costs were adjusted to U.S. dollars, the current time, and the same location. Figure 3 illustrates a

FIG. 3 Cost benchmarks (building cost in U.S. dollars/Guggenheim square meters (GSM)).

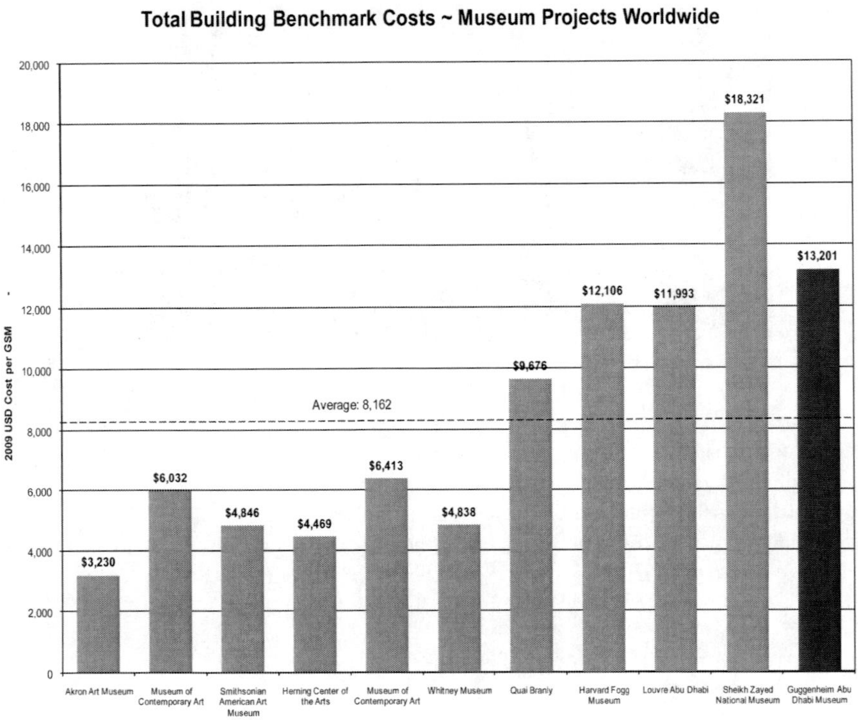

comparison of the Guggenheim cost of over $13,000 per SM, which is much higher than the average cost of the other museums at approximately $8,000 per SM.

Cost Model

To understand the cost of construction for the project, a cost model was prepared prior to the workshop. Each of the cost components were then further broken down in a pareto diagram (highest to lowest cost) and shown for the museum in **Fig. 4**. The team focused on how to lower the costs of the highest components.

Risk Model

A risk model was introduced to help the VM team identify specific risk areas currently observed within the project. Risks were captured in five major categories:

1. management, financial, and administrative risks,
2. environmental, geotechnical risks,
3. technical risks,
4. security risks, and
5. implementation risks.

FIG. 4 Cost model (pareto of museum construction cost).

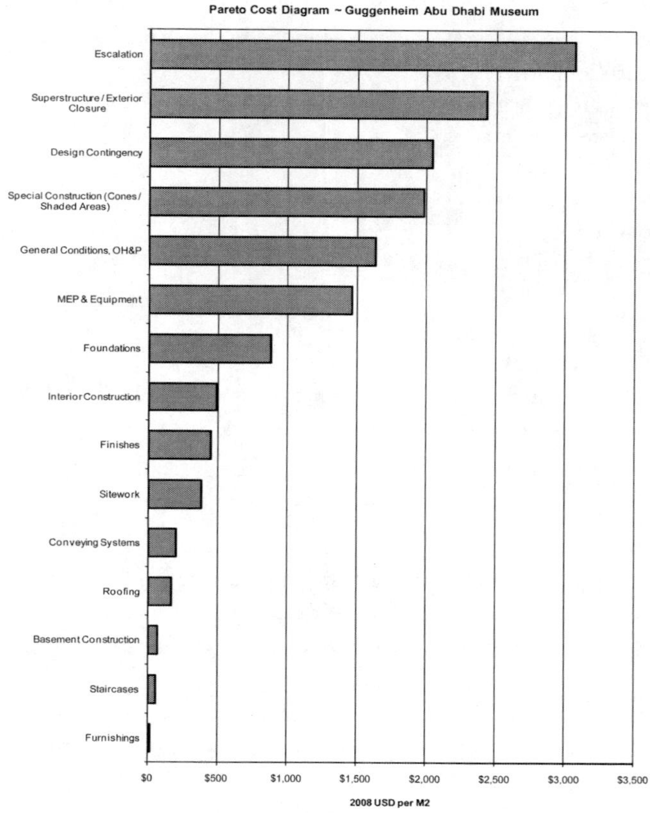

Major risks included the schedule, budget limitations, inflation, inclement weather, new unproven systems like the cones and shards, design approvals and changes, availability of qualified contractors, change orders, field changes, defects or rejected material submittals, and estimating. The VM team reviewed the risk areas and generated contingency plans to minimize "medium and high" risks items associated with the project.

Sustainability Model

The LEED USGBC [11] (Leadership in Energy and Environmental Design) model was prepared by the VM team during the workshop. It helped the VM team identify opportunities to make the project more sustainable. Enough points were identified to achieve "silver" certification.

A second LEED model, the Estidama Pearl rating system [12], was also prepared by the VM team. Suggestions to maximize the sustainability opportunities for this project were captured by the VM team.

Constructability Checklist

In an effort to more fully understand the challenges that this project will produce during the construction phase, a constructability checklist was developed by the VM team during the workshop. It involves identifying the challenges anticipated in several key areas as follows:

- staging and mobilization,
- maintenance considerations,
- site work,
- structure,
- exterior closure,
- roofing, and
- finishes.

Schedule (Time) Model

The design team prepared a design and construction schedule for the VM team review. It was used to review potential opportunities in minimizing or reducing possible risks of meeting the current project schedule.

FAST DIAGRAM

The VM team prepared a function logic diagram to help understand the overall purposes of the new museum. This diagram describes the primary functions of the project that will "enhance Abu Dhabi culture" by "attracting international attention" (with world class architecture) and by "enlightening the public" (with world class art and galleries for viewing and educating visitors), which are considered the basic functions of the project. A portion of the function–logic diagram is shown as **Fig. 5**. Later, the team identified alternative ideas for meeting the functions shown on the diagram. The function analysis served a tool to creatively inspire change in the current design.

TEAM ORGANIZATION AND CREATIVITY

With the large number of participants it was important to organize the teams to maximize their participation and creativity. Two sessions of creativity were organized as illustrated in **Fig. 6**. Session 1 focused on landscape, architectural, and structural in track I, and mechanical and electrical systems in track II. The second session focused on sustainability in track III, and risk and constructability in track IV. The team members were re-mixed to optimize their subject matter expertise from session 1 to session 2.

VM IDEA EVALUATION AND PROPOSAL DEVELOPMENT

Each team did a preliminary evaluation of their own ideas. The most promising were presented to the complete group for their input. Based on these discussions, the most significant ideas were then developed into VM proposals by the small

FIG. 5 FAST diagram of the museum.

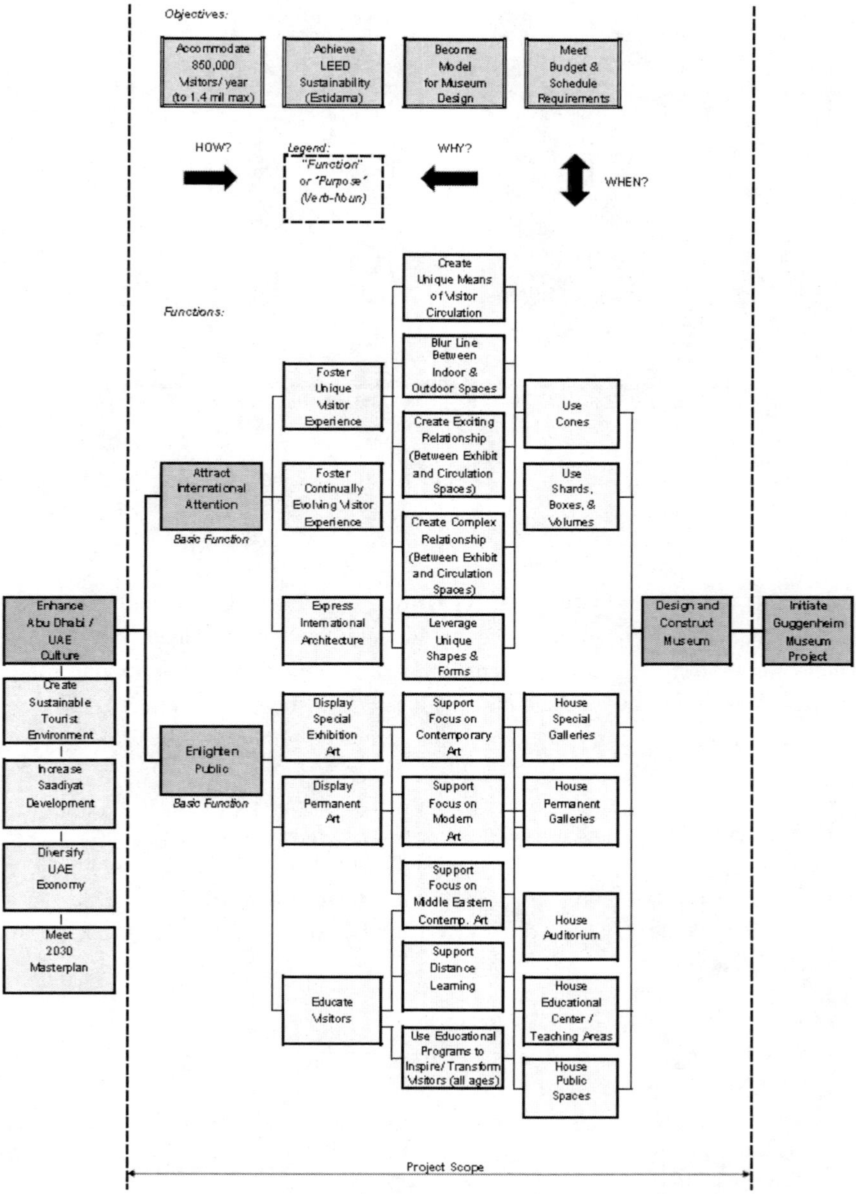

groups. Development included preparing life-cycle cost analysis, engineering calculations, sketches, further research on viability, and ultimately a recommendation listing the advantages of the proposed alternative.

FIG. 6 Two creativity sessions.

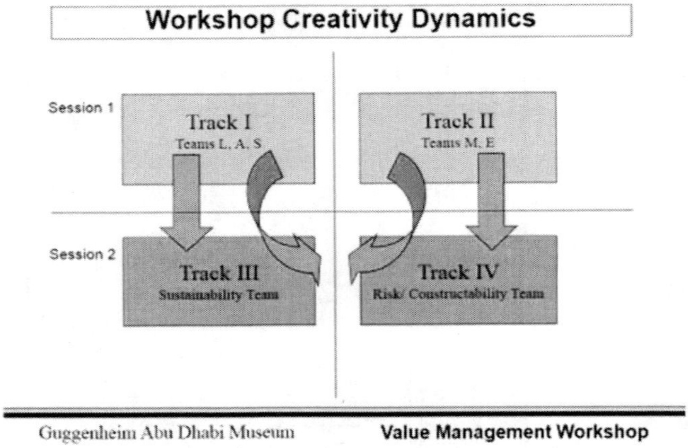

VM Results

VALUE MANAGEMENT RECOMMENDATIONS

The value management team identified 300 ideas for value enhancement. Of these, 81 ideas were selected for development into recommendations for improving the performance and/or lowering the initial and life-cycle cost of the project. "Big" ideas included the following (number refers to the proposal number in the VM report [13]):

- Coordinate park into the master plan as shared resource (L-13).
- Alternatives to address site flows, security, and screening (L-57 and L-68).
- Support original program efficiency (A-5, A-10, A-31, A-85, and A-292).
- Simplification/reduction of space in basement (A-61 and A-293A).
- Create multiple phases for project (A-10 and A-62).
- Accelerate seismic investigation (S-23).
- Enclosing the main atrium space (M-51A).
- Changes to sprinkler strategy (M-150 A-C).
- Air-handling air curtain and redundancy changes (M-158, M-162, and M-265).
- Leveraging desiccant dehumidification with solar regeneration (G-235).
- Lighting strategy changes (G-102, G-112, and G-273).
- Improving museum synergies (R-8).
- Advance the design to pre-purchase long lead items (R-244).

PERFORMANCE IMPROVEMENT

Figure 7 reflects the percentage of VM proposals focused on performance improvements by category.

FIG. 7 Potential performance improvements by category.

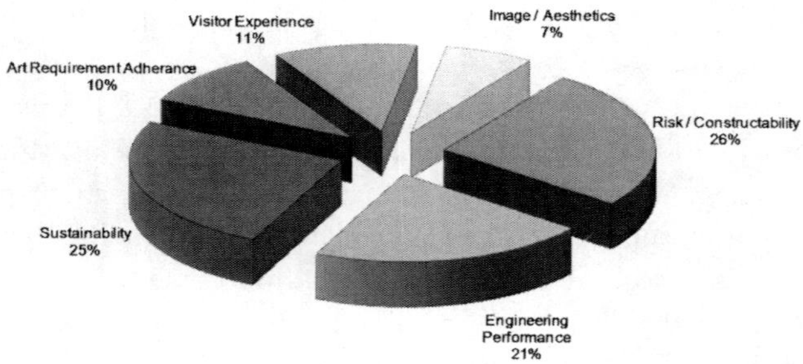

Other benefits include: reduces risks, improves energy performance, leverages new technology, reduces circulation conflicts, simplifies construction, reduces exterior closure, improves egress flows, increased museum flexibility, lightens structure, reduces construction schedule, reduces risks associated with fuel, satisfaction of Estidama or LEED requirements, maintainability, energy, initial cost savings, and life-cycle cost savings.

COST SAVINGS

Some recommendations will generate significant savings for the project while others will add costs. Cost savings identified by the VM team is illustrated in **Fig. 8**. The life-cycle costs were also reduced by a similar percentage. These included energy, maintenance, and replacement costs.

FIG. 8 Potential cost savings.

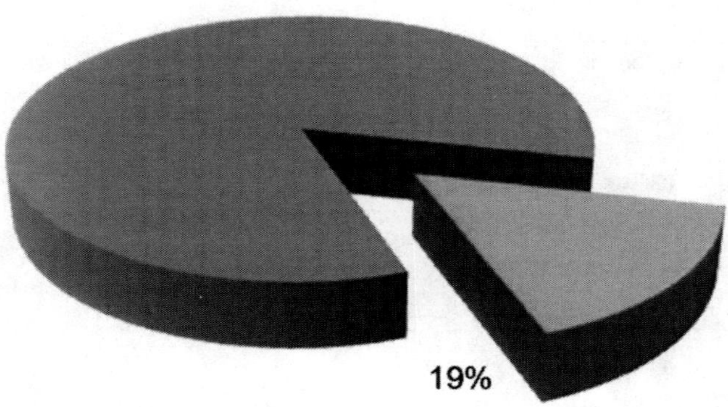

Conclusion

This VM study, performed at the schematic design stage, was a success in part because of the use of function analysis, which helped inspire change. Both performance and life-cycle cost savings resulted in this study, which followed the SAVE value methodology and ASTM Standards E1699 [7] and E2013 [8]. This building example, designed by world-renowned architect Frank Gehry, demonstrates how the value methodology and a team of subject matter experts led by a CVS produces performance improvements even to the most sensitive architectural designs. It also illustrates how much can be achieved in a 3-day VM workshop. At the conclusion of the VM workshop, the design team congratulated all the participants for their thought-provoking ideas and recommendations.

References

[1] SAVE International, "Value Standard and Body of Knowledge," Dayton, OH, 2007.

[2] Office of Management and Budget, "Value Engineering," *Circular A-131*, Washington, D.C., 1993.

[3] U.S. Federal Government, 1996, Public Law 104-450, Washington, D.C.

[4] Miles, L. D., *Techniques of Value Analysis and Engineering*, McGraw Hill, New York, 1961, p. 4.

[5] Miles Value Foundation, "Value Methodology, A Pocket Guide to Reduce Cost and Improve Value through Function Analysis," GOAL/QPC, Salem, NH, 2008.

[6] Kirk, S. J. and Spreckelmeyer, K. F., *Enhancing Value in Design Decisions*, University of Michigan Press, Ann Arbor, MI, 1998.

[7] ASTM E1699-13: Standard Practice for Performing Value Engineering/Value Analysis of Projects, Products and Processes, *Annual Book of ASTM Standards*, ASTM International, West Conshohocken, PA, 2013.

[8] ASTM E2013-12: Standard Practice for Constructing FAST Diagrams and Performing Function Analysis during Value Analysis Study, *Annual Book of ASTM Standards*, ASTM International, West Conshohocken, PA, 2013.

[9] ASTM E917-05: Standard Practice for Measuring Life-Cycle Costs of Buildings and Building Systems, *Annual Book of ASTM Standards*, ASTM International, West Conshohocken, PA, 2010.

[10] Kirk, S. J. and Dell'Isola, A., *Life Cycle Costing for Facilities*, Wiley, Hoboken, NJ, 2003.

[11] USGBC, "Leadership in Energy and Environmental Design (LEED)," U.S. Green Building Council, Washington, D.C., 1994.

[12] Abu Dhabi Urban Planning Council, "Estidama Pearl Rating System," Abu Dhabi, UAE, 2010.

[13] Kirk Associates, "Guggenheim Abu Dhabi Museum VM Study," Seattle, WA, 2011.

STP 1586, 2014 / available online at www.astm.org / doi: 10.1520/STP158620140030

M. Kasi,[1] D. Tyler,[2] and A. Zacharias[3]

Asset Management of Transportation Project at Planning Level

Reference

Kasi, M., Tyler, D., and Zacharias, A., "Asset Management of Transportation Project at Planning Level," *Ideas to Impact: How Building Economic Standards Keep You on Track*, STP 1586, Robert E. Chapman and Michael N. Goodkind, Eds., pp. 35–51, doi:10.1520/STP158620140030, ASTM International, West Conshohocken, PA 2014.[4]

ABSTRACT

The Wisconsin Department of Transportation (WisDOT) successfully conducted a value planning study to manage the asset value of proposed improvements for a corridor project. The proposed work included reconstruction of the existing road section, addition of several miles of a third lane to carry peak traffic, and bridge reconstruction to accommodate the widened roadway. The study used the following standards as a guide in managing the asset: ASTM E1699-13, ASTM E2013-12, and ASTM E917-05. Using the aforementioned ASTM standards, this study addressed the asset management requirements of the Federal Highway Administration (FHWA). Most of the bridges in this corridor will be rehabilitated instead of reconstructed; pavement reconstruction will be minimized. WisDOT accepted several value engineering recommendations and the estimated construction costs were reduced from $\$305 \times 10^6$ to $\$198 \times 10^6$.

Manuscript received March 25, 2014; accepted for publication June 10, 2014; published online August 5, 2014.

[1]Chairman Emeritus, Alfred Benesch & Company, 205 North Michigan Avenue, Suite 2400, Chicago, IL 60601, e-mail: mkasi@benesch.com

[2]WisDOT Bureau of Project Development, 4802 Sheboygan Avenue, Room 651, Madison, WI 53705, e-mail: Daniel.Tyler@dot.wi.gov

[3]Project Manager, Alfred Benesch & Company, 1300 West Canal Street, Suite 150, Milwaukee, WI 53233, e-mail: azacharias@benesch.com

[4]ASTM Symposium on *Ideas to Impact: How Building Economic Standards Keep You on Track* on October 5, 2014 in New Orleans, LA.

Keywords
asset management, function analysis, FAST diagram, value analysis, value engineering

Introduction

The ASTM Building Economic Standards address the big picture of transportation needs. In today's economic conditions, asset management is critical. Project decisions should be based on life-cycle cost that includes first cost, operating cost, and maintenance cost. Funds also should be spent wisely by eliminating unnecessary cost. Asset management, as defined by the Federal Highway Administration (FHWA) [1], addresses these factors. ASTM Economic Standards offer tools and methods to fulfill the requirements.

According to the FHWA,*"An asset management decision making framework needs to be guided by performance goals, cover an extended time horizon, draw from economics as well as engineering, and consider a broad range of assets. At its most basic level, Transportation Asset Management links user expectations for system condition, performance, and availability with system management and investment strategies. Regardless of the definition, the focus is on performance of assets. The underlying goal of asset management is to take a broad approach to resource allocation and programming decisions that will provide greater value to the system and overall satisfaction for end users through improvements in program effectiveness and system performance."*

ASTM Standards outline the value engineering [2] (VE) and life-cycle cost [3] process that satisfy the core principles of asset management. This paper shows examples of how the following three standards shown below help to manage assets of the project:

- ASTM E1699–14: Standard Practice for Performing Value Engineering (VE)/ Value Analysis (VA), of Projects, Products and Processes [2].
- ASTM E2013–12: Standard Practice for Constructing FAST Diagrams and Performing Function Analysis During Value Analysis Study [4].
- ASTM E917–13: Standard Practice for Measuring Life-Cycle Costs of Building and Building Systems [3].

Using the above ASTM standards, this study addressed the asset management requirements of the FHWA.

VALUE ENGINEERING

Function analysis in a VE job plan defines the purpose of proposed improvements, classifies them, and assigns cost to those improvements [5]. The value engineering job plan (see **Fig. 1**) helps the planners to compare other possibilities and get the best value for the effort. The case study that is presented here was performed at the planning level. It is part of Wisconsin DOT's Asset Management approach. The VE team utilized the ASTM standards to set a course for asset management of the corridor.

FIG. 1 VE job plan.

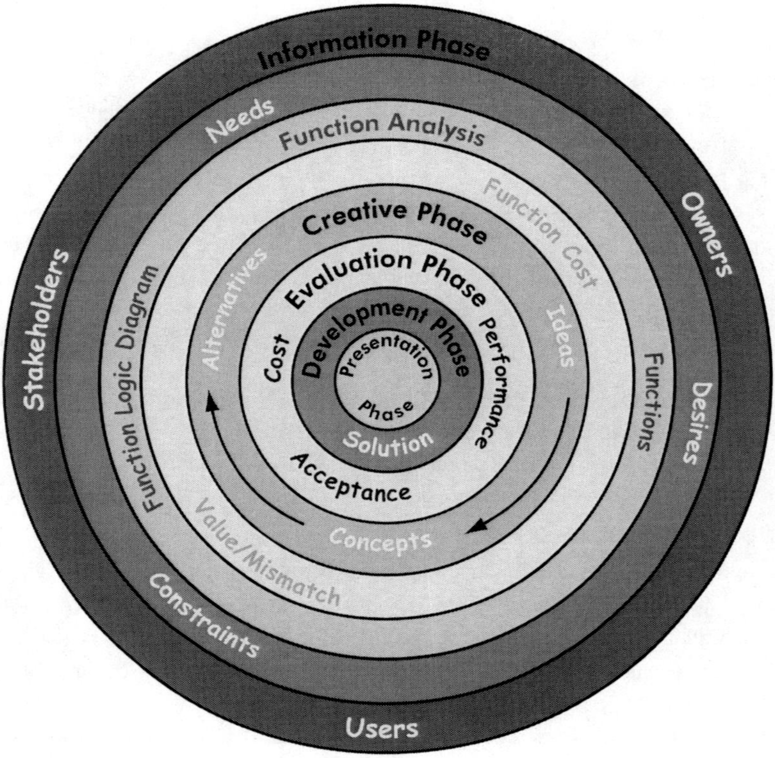

Value engineering is a systematic team approach to creatively enhance the value of a project or product. In the industry, VE is often referred to with various other names such as value analysis, value planning and value management. When the technique is applied during project planning, it is called value planning. If the techniques are used on an already designed or constructed project, it is called value analysis. The approach to each of these varies slightly but the job plan remains the same.

VE advocates a step-by-step, systematic approach to solve a problem. This systematic approach is embodied within a job project description and VE plan that has several phases and imposes a set of rules that must be adhered to. The rules may appear to be simple, but they are vital to the success of the value engineering study. This paper describes how the Wisconsin Department of Transportation (WisDOT) utilized VE [2] to manage assets in the planning phase of the subject project.

A typical VE job plan as outlined in ASTM E1699 [2] consists of the following phases:

- Information phase
- Function analysis phase (ASTM E2013) [4]

- Creative phase
- Evaluation phase
- Development phase
- Presentation phase

INFORMATION PHASE

The purpose of the information phase is to gain an understanding of the project and the stakeholders affected by the project. The information phase can be summarized as follows:

- Review all relevant information on the project, including the project description and scope of work
- Identify the owners, users, and stakeholders
- Identify the needs, desires, and constraints of the owners, users, and stakeholders
- Develop project related functions using the stakeholder needs, desires, and constraints
- Determine the task, basic function(s), and supporting functions of the project
- Estimate the cost of project elements and each critical function
- Analyze the owner's and stakeholder's attitude toward each function

FUNCTION ANALYSIS PHASE

In planning any facility, function classification based on customer needs is the key in delivering the correct product or service [4]. Function analysis helps the planners to properly allocate the funds among the five classifications: basic functions, assure dependability, assure convenience, satisfy stakeholders, and attract stakeholders (see **Fig. 2**).

FIG. 2 Customer function logic diagram [4].

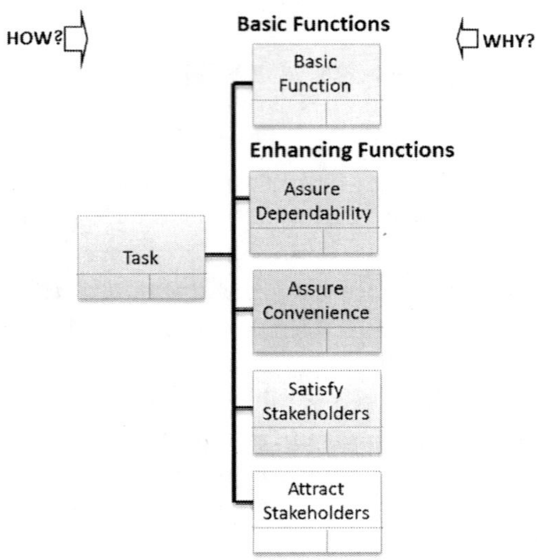

CREATIVE PHASE

The purpose of the creative phase is to identify ideas for improving the project. The ideas are combined into alternatives that will enhance performance or acceptance at a reasonable cost.

EVALUATION PHASE

The purpose of the evaluation phase is to identify the most outstanding alternatives for further development. This identification is accomplished through a series of screening processes that sort the ideas by comparison and combination. Using these ideas, alternatives will be developed. These alternatives will be rated for performance, acceptance and cost.

DEVELOPMENT PHASE

The purpose of the development phase is to add information that will facilitate selection of a preferred alternative. This will be accomplished through a comparison among the remaining alternatives. The following rules should be considered during the Development Phase:

- Recognize alternatives that may be unique
- Conduct research, as required, to provide additional information
- Analyze the weaknesses of the selected alternatives and provide improvements

PRESENTATION PHASE

The purpose of the presentation phase is to convey the intent of the alternatives developed during the study. In addition, it allows the decision makers to interact with the VE team to agree or disagree with the findings.

PROJECT DESCRIPTION

The Wisconsin Department of Transportation (WisDOT) planned to reconstruct and widen the segment of Interstate Highway (IH) 39 between State Trunk Highway (STH 54) and US Highway (USH) 10 West (see **Fig. 3**) over the next 35 years. IH 39 is a four lane divided rural freeway section. The concrete pavement was initially constructed in the 1960s. It was rubblized and overlaid with asphalt in the 1990s. WisDOT is performing analyses of the corridor to detail the exact scope of work. In the concept phase, WisDOT developed a preliminary list of 60 projects involved in this undertaking [6]. The work involves reconstruction of the entire existing pavement section, including ramps, improvement of mainline geometrics, replacement of all bridges on IH 39 except one, improvement of interchange geometrics, additions of a third lane for most of the corridor and addition of a south bound auxiliary lane between County Trunk Highway (CTH) HH and STH 66.

FIG. 3 IH 39.

PROJECT FUNCTION ANALYSIS

The VE Team developed a list of owners, users, and other stakeholders affected by the project and then developed a list of the stakeholders' expectations from the project. These expectations were then considered as functions of the project. Based on the stakeholder preferences, a function analysis system technique (FAST diagram) [4] for the project was developed (see **Fig. 4**).

The function analysis system techniques (FAST) describe the purpose of the improvement of the corridor. The corridor was designed to satisfy the needs of the user's years ago. However, due to pavement and structure conditions and increased traffic, its reliability is lowered. The function of the proposed improvement is to *increase reliability*. The next question is how to *increase reliability*. The answer is to *eliminate deficiencies, which is a basic function*. Basic functions, by definition, are the absolute minimum that needs to be performed.

According to the customer FAST, there are four classifications of enhancing functions that must be explored. They are: (1) assure dependability, (2) assure convenience, (3) satisfy stakeholders, and (4) attract stakeholders. The functions under the "dependability" classification are *safeguard drivers* and *protect structures*. "How do we protect structures?" is answered by two functions: *increase vertical clearance* and *increase horizontal clearance*. These were fulfilled by raising the bridges and replacing superstructures with shallower depths. Another dependability function, *safeguard drivers*, was addressed by three functions: *drain roadway, improve sight distance*, and *allow emergency access*. The function *allow emergency access* is achieved by widening the shoulders to Wisconsin DOT standards.

FIG. 4 Project FAST diagram.

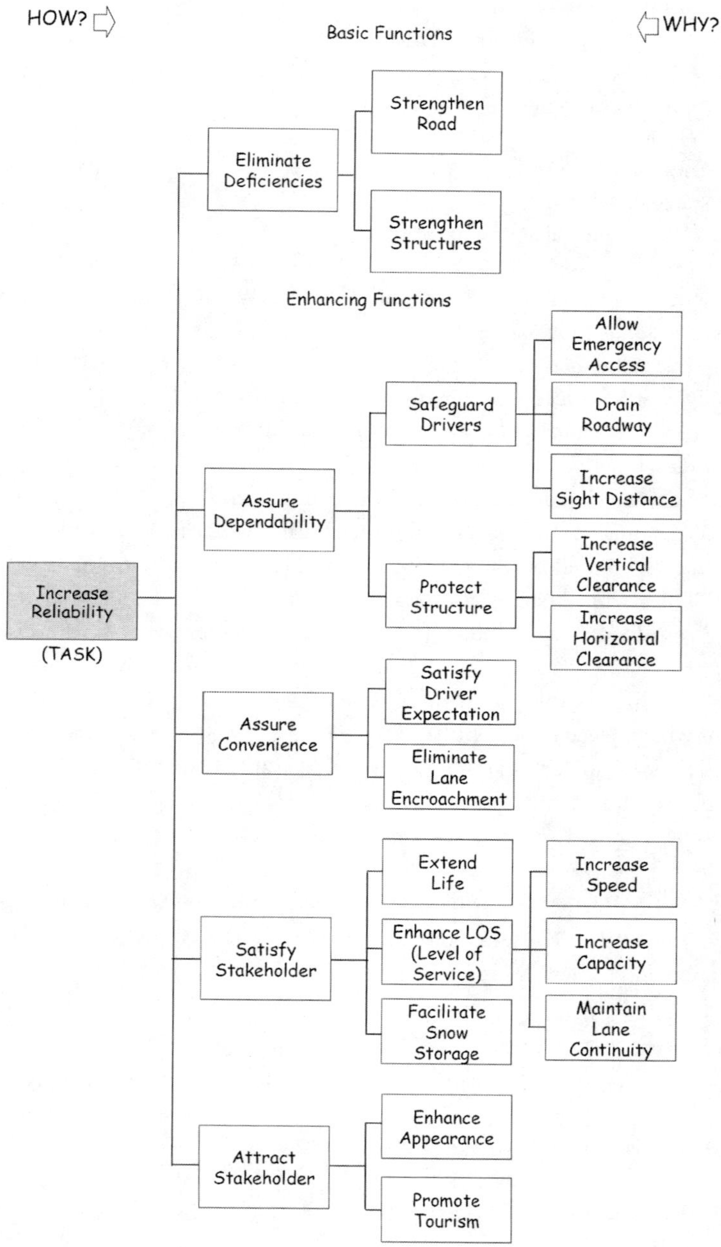

The functions under the "convenient" classification are addressed by the proposed design, meeting the design standards (*satisfy driver expectation*) and limiting the stormwater flow within the shoulders (*eliminate encroachment*).

Under the third classification, "satisfy stakeholders," the VE team placed the function *enhance level of service* (LOS).

LOS is measured in a scale of A to E [7]:

- A: Free flow
- B: Reasonable free flow
- C: Stable flow
- D: Approach unstable flow, operating at capacity
- E: Unstable flow, stop and go traffic

The function *enhance LOS* can be placed under any one of the four categories. If the existing LOS is E or worse, then the function will be changed to *improve LOS* and will be placed under *assure dependability*. If the LOS is either C or D, then *improve LOS* will be placed under assure convenience. If the LOS is either B or A, the function will be *enhance LOS* and will be placed under *satisfy stakeholders* or *attract stakeholders*.

In addition, the classification changes are based on the type of traffic. The Wisconsin DOT Manual distinguishes commuter traffic (weekdays) from recreational traffic (weekends). Based on the guidelines, *Improving commuter traffic* is either a dependability or convenience classification. *Improving recreational traffic* falls under *satisfy stakeholders*. Another function, *facilitate snow storage*, can also be placed in three categories. If it is a two-lane highway, any snow stored would impede the traffic, and be placed under *assure dependability*. If it is a four-lane highway, it will be placed under *assure convenience* and under *satisfy stakeholder*. If it is a six-lane or four-lane highway, recreational traffic controls the LOS. In this case, it is placed under *satisfy stakeholder* functions.

The FAST diagram [4], as shown in **Fig. 4**, reflects these functions and their classifications.

The VE team developed approximately 160 ideas that could potentially enhance the value of the project. The ideas were then screened into nine proposals, including one validation and 14 design suggestions.

VE PROPOSALS

This paper discusses the following three proposals out of the nine proposals developed:

1. Addition of a third lane ($43.1 m)
2. Replacement of 32 bridges ($39.2 m)
3. Replacement of pavement ($69.63 m)

PROPOSAL 1: CAPACITY OF ROADWAY

In the information phase, the team reviewed the reason for the third lane addition; it is based on the traffic demand at the 30th hour volume. WisDOT policy is to use the 30th highest hour volume (K30) as the design hour volume for mainline freeways, mainline multilane highways, and rural two-lane facilities and for urban arterial analysis of through movements. There may be unique circumstances where K30 is not realistic to use because of exceptionally high hourly volume peaking characteristics. These conditions may occur on highly recreational routes, or routes that are in close proximity to a stadium or seasonal shopping mall. Additionally, higher design hour volumes may be justified when the LOS using K30 cannot be achieved because of social, environmental, or financial constraints. When higher design volumes are justified, the LOS evaluation should also consider the 100th highest design hour for rural or small to medium urban areas and 250th highest hour for highly urbanized areas (>200 000 population) with heavy daily traffic. The FHWA must approve deviations from the K30 design hour on interstate projects [8].

It was planned to add a third lane as part of the improvement. The cost of adding this third lane is approximately $\$43.1 \times 10^6$, as shown in **Table 1**.

TABLE 1 Added cost to construct a third lane.

Description	Major Roadway Items	Major Structures	Total Cost
SB Mainline—STH 66 to Old USH 10 E	$1 210 544	$1 183 000	$3 185 183
NB Mainline—STH 66 to Old USH 10 E	$1 210 544	$4 158 000	$6 160 183
SB Mainline—Old USH 10 E to CTH HH	$2 315 492	0	$3 420 382
SB Mainline—BUS 51 to STH 66	$1 948 168	0	$3 683 915
SB Mainline—Casimir Road to BUS 51	$823 542	0	$1 216 514
SB Mainline—USH 10 W to Casimir Road	$3 391 289	0	$5 009 520
NB Mainline—BUS 51 to Casimir Road	$2 815 200	0	$4 158 537
SB Mainline—CTH HH to CTH B	$2 915 363	0	$4 513 209
NB Mainline—STH 66 to BUS 51	$2 307 938	0	$3 409 224
NB Mainline—Casimir Road to USH 10 W	$1 562 333	0	$2 307 836
NB Mainline—CTH HH to Old USH 10 E	$1 241 544	$4 158 000	$6 080 006
TOTAL			**$43 144 572**

Roadway improvements, in general, have two parts: capacity of the road and condition of the road. Strengthening the road is a basic function. This can be accomplished by rehabbing the structure with the proper maintenance program. Any additional improvement including reconstruction will be based on the function *extend life* which is a *satisfy stakeholder* function. The capacity of roadway is based on road use. Some states do not differentiate the importance of week day traffic and weekend traffic. Some states give higher preference to week day traffic since they impact business hours. It also depends upon the states that rely on tourism. In such cases, the states may give higher priority for recreational traffic. Function analysis communicates this importance by placing them under the proper category.

If the weekday traffic is given the same priority as the weekend traffic, the cost of adding an additional lane is classified in **Table 2**.

TABLE 2 Function classification when weekday and weekend traffic are given the same priority.

Function Classification	Weekday or Weekend Traffic
Assure dependability	If the LOS is E or F
Assure convenience	If the LOS is C or D
Satisfy stakeholders	If the LOS is B or A
Attract stakeholders	

ASTM E2013-12 [4], section 6.6.13 states "*Compare function cost and function preference. Table 1 shows four possible combinations of cost/preference. If the cost of a function is high and the stakeholder thinks its importance is low, the result is a mismatch (Type A in Table 3)*. On the other hand, if the cost of a function is low and the stakeholder rates its importance as high, a high value is achieved and the stakeholder has a match (Type D in **Table 3**). These are the two extremes of the cost/preference measurement."

TABLE 3 Illustration of cost/preference combinations.

Type of Combination	Function Cost	Function Preference	Function Value
A	High	Low	Mismatch (Mandatory Value Analysis) (VA)
B	High	High	Candidate for VA
C	Low	Low	—
D	Low	High	Match

In a recent VE of a similar situation for another agency, the team decided to place the cost of *enhance LOS* under *assure dependability*. This is due to the decision of the agency to give a higher preference to weekend traffic. This means that the need is high and the cost is high (see **Fig. 5**). This led to the decision to add a third lane.

FIG. 5 Analysis of adding the third lane.

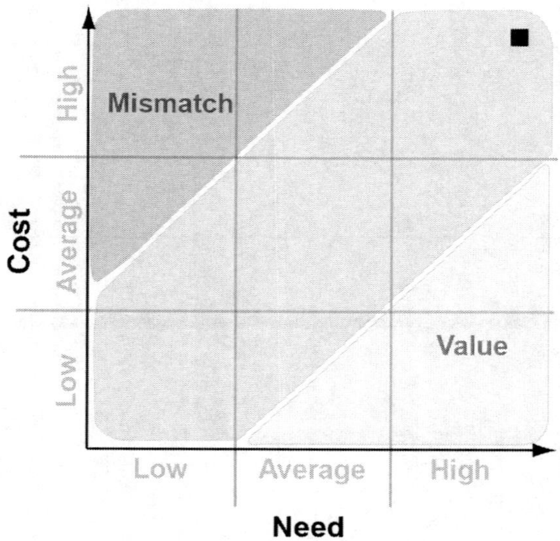

As stated earlier, the WisDOT manual allows the level of service based on the 100th hour peak volume to be used for recreational traffic. If the weekday traffic is given priority over the weekend traffic, the classification is as shown in **Table 4**.

TABLE 4 Function classification when weekday traffic is given priority.

Function Classification	Week Day Traffic	Weekend Traffic
Assure dependability	If the LOS is E or F	
Assure convenience	If the LOS is C or D	If the LOS is E or F
Satisfy stakeholders	If the LOS is B or A	If the LOS is C or D
Attract stakeholders		If the LOS is B or A

However, by using a lower priority in computing the LOS for the recreational traffic, adding a third lane is a *satisfy stakeholder* function. See **Table 5**.

In this case, the need to satisfy the 30th hour peak is low and adding the cost of adding a third lane is high, which is a mismatch. The need to satisfy the 100th hour is high and the cost is low; this puts it under value zone (see **Fig. 6**).

TABLE 5 Relationship of function classification to need.

Function Classification		
Assure dependability	Based on 100th hour peak volume (K100) Level of Service C	Existing 2 lane is sufficient
Assure convenience		
Satisfy stakeholders	Based on 30th hour peak volume (K30) Level of Service F	Need a third lane
Attract stakeholders		

FIG. 6 Analysis of traffic impact.

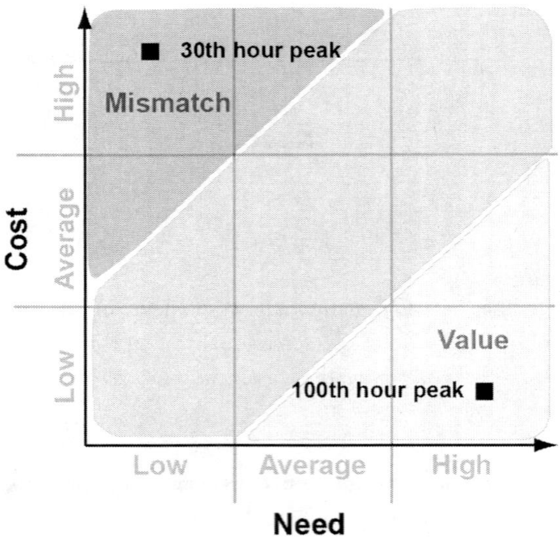

The decision was made to adopt the 100th hour peak volume and not add a third lane. The cost of adding a third lane was to satisfy the function *enhance LOS*. By placing it as a *satisfy stakeholder* function, it was decided to eliminate the third lane. This allows the project to avoid approximately $\$43.14 \times 10^6$ in project costs.

PROPOSAL 2: BRIDGE STRENGTHENING

The second proposal is the replacement of bridges that are structurally inadequate. All of the bridges along IH 39 were built in the late 1960 s. WisDOT has a very structured maintenance program that is based on the approach to life-cycle cost.

The inspection reports indicate bridge sufficiency ratings between 80 and 95. Sufficiency ratings, used by FHWA, are intended to indicate a measure of its ability to remain as a safe structure. Bridge inspectors, during biannual inspections, calculate the ratings from 0 to 100. The FHWA suggests that a bridge should be replaced if the sufficiency rating is below 50 and rehabilitated when it is below 80. If the sufficiency ratings are between 50 and 80, the structure needs rehabilitation. In such a case, engineers rely on the National Bridge Inventory (NBI) [9]. This includes a structural evaluation of the deck, superstructure and substructure on a scale of 0 to 9. The scale is defined below:

- 9: Higher than needed
- 4: Meets minimum
- 0: Close the bridge

Figure 7 shows the flow diagram of the bridge management program.

FIG. 7 Bridge management program.

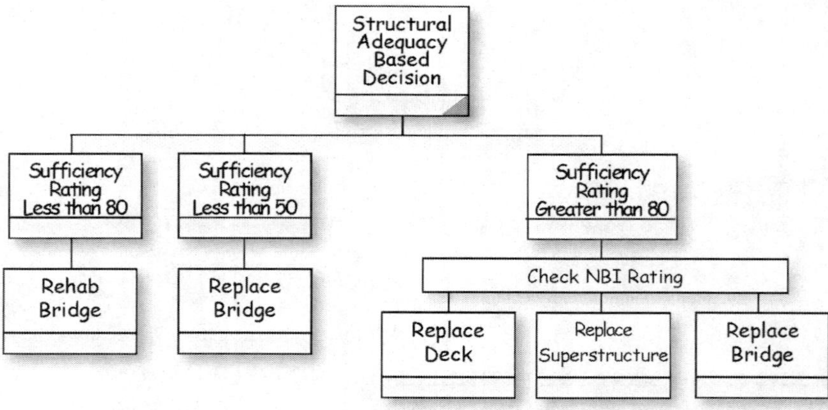

Most of the bridges have a sufficiency rating of 90 or higher. Most of the decks have an NBI rating of 6 or better. This is the result of a very good maintenance program by WisDOT. Consideration should be given to utilize a reasonable life of all good bridges. This will reduce the number of bridges to be replaced. This can defer the replacement cost of $\$16.4 \times 10^6$. **Table 6** shows the existing bridges, their ratings, cost of replacement, and cost of rehabilitation. The cost difference between satisfying the above requirements ($\$22.86 \times 10^6$) and total replacement ($\$39.23 \times 10^6$) is $\$16.38 \times 10^6$. The cost of *assure dependability* is $\$22.86 \times 10^6$. The additional cost of $\$16.38 \times 10^6$ is to *satisfy stakeholder* classification. The VE team recommended the elimination of cost ($16.38 M) to *satisfy stakeholder* function.

TABLE 6 Existing bridges, their ratings, cost of replacement, and cost of rehabilitation.

Feature	Sufficiency Rating	Clearance (FT)	NBI Rating			As Given - Replace		VE Alternative - Rehab	
			Deck	Super	Sub	Structure Cost	Special Constr.	Structure Cost	Special Constr.
STH 54 over IH 39	97.0	16.17	7	7	6	$2,068,000	$0	$921,675	$0
IH 39 over Br Wisc. River (Culvert)	92.1	N/A	7	8	8				
NB IH 39 over CTH B	94.6	14.11	6	7	6	$1,628,000	$263,000	$729,750	$0
SB IH 39 over CTH B	83.2	14.11	6	5	6	$1,625,000	$263,000	$728,250	$0
NB IH 39 over RR	94.2	N/A	6	7	6	$1,894,000	$245,000	$1,894,000	$245,000
SB IH 39 over RR	95.0	N/A	6	7	7	$1,807,000	$245,000	$1,807,000	$245,000
NB IH 39 over Little Plover River	94.8	N/A	7	7	7	$641,000	$105,000	$287,625	$0
SB IH 39 over Little Plover River	83.8	N/A	6	5	8	$641,000	$105,000	$287,625	$0
Porter Road over IH 39	96.6	17.58	9	9	9	$212,000	$0	$212,000	$0
CTH HH over IH 39	96.0	16.17	7	7	7	$3,721,000	$0	$1,667,700	$0
NB IH 39 over CN Railroad	97.3	N/A	7	7	7	$1,650,000	$498,000	$618,000	$0
SB IH 39 over CN Railroad	97.3	N/A	6	6	7	$1,979,000	$498,000	$742,350	$0
USH 10 East/STH 66 West over IH 39	81.6	16.14	6	6	5	$4,562,000	$0	$2,352,525	$0
NB IH 39 over Plover River	94.5	N/A	6	6	7	$1,650,000	$180,000	$836,700	$0
SB IH 39 over Plover River	94.5	N/A	7	7	7	$1,224,000	$180,000	$549,450	$0
Barbaras Lane over IH 39	62.2	16.37	5	8	7	$942,000	$0	$484,200	$0
NB IH 39 over Stanley St/STH 66 East	79.4	14.83	6	5	6	$2,388,000	$255,000	$1,433,800	$0
SB IH 39 over Stanley St/STH 66 East	91.6	14.83	6	6	6	$2,388,000	$0	$1,433,800	$0
Wilshire Blvd over IH 39	91.2	16.24	5	8	6	$733,000	$0	$733,000	$0
SB IH 39 over Reserve St	81.8	14.40	5	5	7	$1,366,000	$218,000	$1,366,000	$218,000
NB IH 39 over Reserve St	85.0	14.67	5	5	7	$1,524,000	$218,000	$1,524,000	$218,000
Business 51 over IH 39	98.0	17.34	7	8	7	$220,000	$0	$220,000	$0
N 2nd St over IH 39	76.2	16.17	5	8	7	$963,000	$0	$963,000	$0
IH 39 over Hay Meadow Creek (Culvert)	75.6	N/A	7	6	8				
Casmir Road over IH 39	98.1	16.84	7	7	7	$132,000	$0	$132,000	$0
CTH X over IH 39	96.9	16.30	7	8	7				
						$35,958,000	$3,273,000	$21,924,450	$926,000
						$39,231,000		$22,850,450	
					Cost Difference	$16,380,550			

PROPOSAL 3: PAVEMENT STRENGTHENING

The third proposal is the proposed pavement reconstruction. IH 39 is a four-lane divided rural freeway section. The concrete pavement was constructed in the 1960 s. The concrete pavement was rubblized and overlaid with approximately five inches (12.70 cm) of hot mix asphalt (HMA) in 1998. Since that time, tenting of the overlay has occurred during the winter months. In 2012, a segment was rehabilitated by milling 0.75 in. (1.90 cm) and adding a 2.00 in. (5.08 cm) HMA overlay. This work also includes a second year rout and seal of the new HMA overlay.

It was considered to entirely reconstruct the existing pavement over the next several years. This will involve removing the existing pavement structure, including the rubblized layer, and constructing a new, full-depth concrete pavement. In addition, the as given plan envisioned an overlay of the entire pavement in 2035.

The base construction cost estimate for the reconstruction of the existing pavement is $69 630 000 in 2011 dollars (see **Table 7**). This cost does not include the possible overlay in 2035.

TABLE 7 Cost of reconstruction of pavement.

Description	Year	Project Cost
South project limit to Porter Road	2015	$\$27.44 \times 10^6$
USH 10 East to North project limit	2016	$\$25.24 \times 10^6$
Porter Road to USH 10 East	2019	$\$16.95 \times 10^6$
TOTAL		**$\$69.63 \times 10^6$**

The VE team recommended developing a Pavement Management Program instead of reconstructing the pavement. This would involve continuing the current IH 39 pavement rehabilitation plan of mill and overlay in 2018 and again in 2024. Then, in 2030, the entire asphalt pavement (only) would be replaced with new full-depth asphalt. The rubblized concrete pavement would not be removed in 2030. This VE Alterative would utilize the entire service life of the existing pavement, while distributing expenditures over a longer time period [10].

The comparable base construction cost estimate for this VE alternative considering the mainline pavement only is $32 180 000 in 2011 dollars, as shown in **Table 8.**

TABLE 8 Cost of rehabilitation of pavement.

Description	Year	Project Cost
Pavement mill and overlay	2018	$\$6.46 \times 10^6$
Pavement mill and overlay	2024	$\$6.46 \times 10^6$
Asphalt pavement replacement	2030	$\$19.26 \times 10^6$
TOTAL		**$\$32.18 \times 10^6$**

Rehabbing the pavement at a cost of $\$32.18 \times 10^6$ is a basic function (*strengthen road*) whereas the additional cost of $\$37.4 \times 10^6$ is to satisfy the *extend life* function. It was decided to rehab the pavement and avoid an estimated $\$37.4 \times 10^6$ in project costs.

Conclusion

In conclusion, the study was able to demonstrate that the program cost can be reduced substantially without significantly affecting the value. The recommendations are based on balancing the initial cost, operating cost and maintenance cost. The three proposals resulted in $\$96.92 \times 10^6$ ($\$43.14 \times 10^6$, $\$16.38 \times 10^6$, and

37.4×10^6) of cost avoidance. Value engineering recommendations resulted in reducing the estimated construction cost from 305×10^6 to 198×10^6.

This study confirmed the purpose as stated in ASTM E1699 [2]: "*Use of this practice increases performance; maintains/ improves schedule; maintains/enhances quality within budget; provides a more efficient use of resources; reduces risk; and enhances safety or effectiveness, or both. Perform VE/VA as early as possible in the life cycle of the focus of study, and anytime conditions change, to allow greatest flexibility and effectiveness of any recommended changes. However, VE/VA may be performed at any time during the planning, design, and implementation phases of a project, product, or process.*"

Asset management is practiced by placing higher importance to the basic function *assure dependability* and lesser importance to *satisfy stakeholders*. Improvements based on *assure dependability* functions are cost effective and improve assets. Improvements based on *satisfy stakeholder* functions, in most cases, are not as cost effective. They can be adopted if the budget allows.

ASTM standards positively impact project management by helping deliver an efficient and economical project. They not only help an economical design but also a friendly asset management plan that includes preserving existing structures and roads.

References

[1] FHWA, 2013, "Asset Management," http://www.fhwa.dot.gov/asset/pubs.cfm (Last accessed 13 Oct 2013).

[2] ASTM E1699-13: Standard Practice for Performing Value Engineering (VE) /Value Analysis (VA), of Projects, Products and Processes, *Annual Book of ASTM Standards*, ASTM International, West Conshohocken, PA, 2013.

[3] ASTM E917-05 (2010): Practice for Measuring Life-Cycle Costs of Buildings and Building Systems, *Annual Book of ASTM Standards*, ASTM International, West Conshohocken, PA, 2012.

[4] ASTM E2013-12: Practice for Constructing FAST Diagrams and Performing Function Analysis During Value Analysis Study, *Annual Book of ASTM Standards*, ASTM International, West Conshohocken, PA, 2012.

[5] Kasi, M., *Function Approach to Transportation Projects—A Guide to Value Engineering*, iUniverse, Bloomington, IN, 2009.

[6] Goodkind, M., Kasi, M. and Zacharias, A., "North Junction USH 10 Value Engineering Report," *IH 39, STH 54*, Organization, Location, 2013.

[7] American Association of State Highway and Transportation Officials, *A Policy on Geometric Design of Highways and Structures*, AASHTO, Washington, D.C., 2011.

[8] Wisconsin Department of Transportation, 2013, "Wisconsin Department of Transportation Facilities Development Manual, 11-5, 3.5.1," http://roadwaystandards.dot.wi.gov/standards/fdm/index.htm (Last accessed 16 Oct. 2013).

[9] FHWA, 2013, "Bridges and Structures, National Bridge Inventory," http://www.fhwa.dot.gov/bridge/nbi.cfm (Last accessed 16 Oct. 2013).

[10] Federal Highway Administration, "Life-Cycle Cost Analysis in Pavement Design - In Search of Better Investment Decisions," *FHWA-SA-98-079*, FHWA, Washington, D.C., 1998.

STP 1586, 2014 / available online at www.astm.org / doi: 10.1520/STP158620140034

Joseph L. N. Martin[1]

Classification of Construction Costs—An International Overview from a UK Perspective

Reference

Martin, Joseph L. N., "Classification of Construction Costs—An International Overview from a UK Perspective," *Ideas to Impact: How Building Economic Standards Keep You on Track*, STP 1586, Robert E. Chapman and Michael N. Goodkind, Eds., pp. 52–79, doi:10.1520/STP158620140034, ASTM International, West Conshohocken, PA 2014.[2]

ABSTRACT

The development of elemental cost planning in the UK gave rise to a generally adopted Royal Institution of Chartered Surveyors (RICS) standard for classifying costs for buildings. The existence of data in this format has underpinned the development of cost control and, recently, the UK government"s cost reduction strategy, which is based on the existence of reliable benchmark data. Similar concepts and classifications have also been developed around the world. Attempts to exchange data with other European countries in the Conseil European des Economistes de la Construction (CEEC) identified some significant differences in national usages for similar concepts (i.e., that in our "common" exchange language of cost per unit area for a building type, neither the costs nor the area were measuring the same thing). This resulted in the CEEC Code of Measurement for Cost Planning, which proposes a common grouping for the building elements and identifies the different area definitions. UNIFORMATII will be mapped to the Code. RICS has undertaken surveys to identify element classifications in other countries and produced Principles of Element Classification for Buildings (International) to promote the concept and use of elements. Until recently, there has been no widespread adoption of these methodologies in civil engineering, despite the Building Cost Information Service (BCIS) having produced a Standard Form of Civil Engineering Cost Analysis for use by the UK Defence Infrastructure Organisation (DIO). The government"s

Manuscript received April 1, 2014; accepted for publication June 17, 2014; published online July 14, 2014.
[1]Executive Director, BCIS, RICS, Parliament Square, London, SW1P 3AD, UK, e-mail: jmartin@bcis.co.uk
[2]ASTM Symposium on *Ideas to Impact: How Building Economic Standards Keep You on Track,* on October 5, 2014 in New Orleans, LA.

desire for benchmarks has led to a wider interest in consistent cost breakdown structures for civil engineering and the development of Building Information Modelling (BIM) is premised on the integration of classification systems used by all parties to the development of a project. The revisions to ISO 12006-2 Building Construction—Organisation of Information About Construction Works—Framework for Classification need to be reviewed in this context

Keywords

construction classification, cost planning, elements, entities, benchmarking, building information modelling, measurement, building, civil engineering

Introduction

BCIS AND THE RICS

BCIS is the Building Cost Information Service of the Royal Institution of Chartered Surveyors (RICS).

BCIS is the UK's leading provider of cost information to the construction industry and rebuilding cost data to the insurance industry. For the past 50 years, BCIS has been collecting, aggregating, analysing, modelling, and interpreting data on behalf of the industry. Today, BCIS makes that information easily accessible through online applications, software services, calculators, publications, and data licenses.

BCIS also provides consultancy and research support to clients from both the public and private sectors. Services include measuring price movement, benchmarking, market research, statistical analysis, forecasting, and impact studies. At the core of these services is BCIS' ability and experience in collecting, modelling, analyzing, and interpreting data.

RICS is a global property professional body which aims to:

- Regulate and promote the profession
- Maintain the highest educational and professional standards
- Protect clients and consumers through a strict code of ethics
- Provide impartial advice, analysis and guidance

RICS has 100 000 chartered members (FRICS and MRICS) and globally, 55 000 members in other categories of membership (TechRICS, trainees, and students). It represents, regulates, and promotes the work of these property professionals throughout 146 countries.

RICS members are involved in the whole lifecycle of property, from land management and measurement, through planning, environmental impact assessment and investment appraisal, to managing the construction process and advising on the most efficient use of buildings.

RICS is governed by a Royal Charter approved by Parliament, which requires it to act in the public interest. It is also a professional regulatory body approved by Government (HM Treasury).

WHY ELEMENTS?

It may be said—with about as much exaggeration as may be normally attributed to such statements, but no more—that the publication of the Ministry of Education *Building Bulletin No 4* [1] in 1951 introduced the concept of elemental cost planning to the UK construction industry.

The preface to the bulletin says that with the need for a great number of new schools, it is "essential to explore every possible means of reducing the cost, increasing the speed of school building whilst maintaining standards of quality and educational efficiency." Does that sound familiar? Our drive today to produce better buildings for less money is the same as it ever was.

The bulletin described a "new approach towards a costing technique for use by architects and surveyors." The fact that the quantity surveying profession made the technique their own may explain the changing influence of the two professions over the past half century.

The bulletin distinguishes two aspects of the technique:

"**Cost analysis** aims at examining the cost of schools already planned or built for which bills of quantities and tenders are available; it is in the nature of a "post-mortem."

"**Cost planning** uses the information gained by cost analysis to maintain a surer control over costs of future projects."

The key to this innovation was the concept of Elements. Again, it is worth quoting from the bulletin:

"An architect tends to think in terms of functions and of the means by which he can perform those functions. For example, he sees as one function the exclusion of rain and weather and he looks to a roof to perform this task. It is, for the purposes of cost analysis, immaterial to him whether the roof is of timber and tiles or of concrete and asphalt. He is primarily concerned to know how much it has cost to roof in the building."

"Thus, where the estimator builds up his tender by adding together a large number of relatively small items, classified by trades, cost analysis must reverse this process and break down the tender into groups of material and labour classified according to the functions they perform. These groups have been described as Elements."

Thus an Element is defined as "a part of a building that fulfils a specific function or functions irrespective of its design, specification or construction."

The basis of cost planning is moving from "costing a design" to "designing to a cost." By establishing benchmark costs for Elements for buildings of different functions (e.g., a primary school), a robust elemental cost plan for a future project can be produced at business case stage, based solely on clients" space requirements, that sets the elemental budget for the design team.

What should be stressed is that functional elements were discovered not invented. They have proved to be such a useful idea because they are part of the intuitive design process. It is not necessary to understand the theory behind functional elements to use Elements; however, it may be helpful to explain how they arise.

It starts with a simple question: "what is a building?" Most dictionary definitions use examples, such as "permanent structure, e.g., schools, houses, etc." However, "school" and "house" define the use to which a building is to be put—its "user function." If we define a building"s function—"its object function," i.e., what all buildings do as distinct from the user function, which describes what is done in them, the definition would be "structurally sound, enclosed, usable floor space." When designing a building, therefore, one must design building parts that fulfil these functions, i.e., parts to make it structurally sound, to enclose it, to provide floor space, and to make it useable. Further expansion of "Usable" would include accessible, environmentally controlled, healthy, safe, etc. **Figure 1** shows a breakdown of some of these functions into parts of a building that are recognisable as Elements.

Since all buildings potentially have these Elements, the design process is one of designing the Elements so that they provide a best fit to the users" functional requirements. It is because cost planning and value engineering techniques follow this design process that Elements can be used to optimise design solutions to clients" requirements.

The cost planning technique has served the profession well in offering value added services to clients and the concept of Elements has been incorporated into the development of cost modelling, life cycle costing and value management, and new methods of procurement.

FIG. 1 Elements derived from the functional decomposition of building.

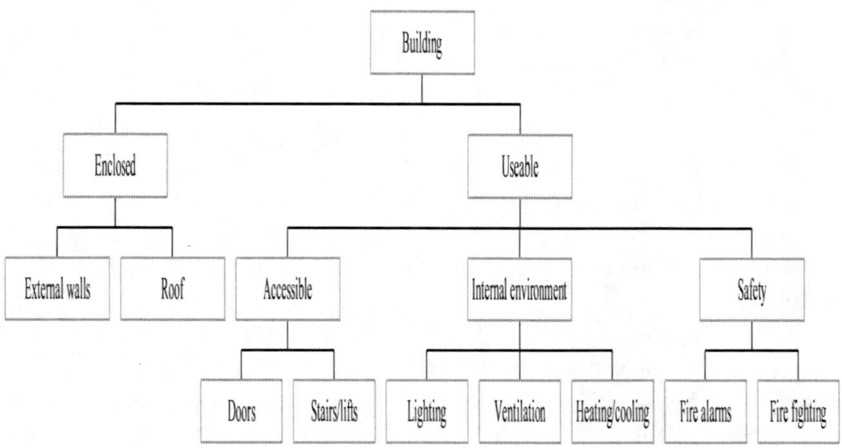

History of UK Standard

To develop a database of costs to facilitate elemental cost planning required standard rules for classifying Elements.

The RICS published its first edition of the *Standard Form of Cost Analysis* (SFCA) in 1961.

In the same year, the Institution set up the BCIS to act as a central repository for the exchange of elemental cost analyses, a role it still fulfils today.

A second edition of the *Standard Form of Cost Analysis* was issued in 1969 [2]. This could be considered the definitive edition as it remained unchanged until 2008 and subsequent editions have only sought to clarify the definitions, expand on the sub-Elements in the light of modern construction techniques, and to take account of questions of interpretation raised by users.

The SFCA is more than a classification of Elements as it contains rules for allocating costs to the Elements to ensure consistency and this requires a compromise between the definition of an Element and what is considered practical. For example, one of the major differences between elemental classifications around the world is the inclusion, or not, of a "Frame" Element. In some classifications the members of a frame are allocated to Elements for "External walls," "Floors", "Roof," etc. However, in the UK the procurement of steel frames makes this allocation difficult; thus, whatever the technical argument might be for including an Element for frame, UK procurement practice dictated its inclusion in the SFCA.

From its inception, BCIS has been guardian of the SFCA and undertook regular reviews with users; however, until 2008, the view was that where the original drafting committee had decided on an issue, this should not be reversed if it was felt that at another review the decision might be reversed.

The 1969 edition [2] was endorsed by all the major government procurement departments and many of them produced their own in-house branded editions for use by their staff. Over the subsequent 45 years, the SFCA data structure has become the accepted default structure for expressing costs for budgeting, estimating, cost planning, cost recording, and benchmarking.

As procurement methods have changed, the SFCA is widely used for structuring Contract Sum Analyses on Design and Build contracts, budgets on two stage tenders and Target Costs on target cost contracts. Many contractors' directors will review their tenders in elemental form as it allows them to interrogate the design and estimating team in a more meaningful way than reviewing a trade or packaged based estimate.

In the UK, "traditional" procurement is based on contractors pricing a Bill of Quantities prepared on behalf of the client from a design also prepared on behalf of the client. The risk of any changes in design, scope, or measurement of the contract rests with the client. On Design and Build contracts, the contractor takes the responsibility for the design and therefore the quantum of the works.

The use of Elements for structuring Bills of Quantities has changed over time. In the 1970s, the vast majority were elemental based, but this was not popular with contractors and its use dwindled; however, in recent years, it has made a comeback and a review of Bills recently submitted to BCIS suggests that approximately 40 % were structured elementally.

The SFCA Elemental structure has also been used for performance specifications.

In 2007, RICS undertook a review of its measurement standards and identified a need for rules for cost planning as the design developed, i.e., rules for measuring designed Elements (i.e. the design solution for an element on a particular scheme, defined in ISO 12006.2:2001 [3] as an element for which the work result(s) have been defined), sub-Elements and components. Thus, while there were rules in the SFCA for measuring undesigned Elements, i.e., irrespective of its design, specification or construction, and in the RICS Standard Method of Measurement [4] for measuring trade work results for inclusion in a Bill of Quantities, there were no agreed rules for measuring the whole of a designed Element. For example, the SFCA rule for measuring External Walls as an Element is to measure the area on the internal face so that the measurement is the same irrespective of the thickness of the designed wall, and the SMM rule for measuring the brick external skin of a cavity wall is to measure along the centre line of the wall. There was no rule for measuring a complete cavity wall including the internal and external skin, cavity, insulation, etc.

This resulted in the publication of a new 3rd Edition of the SFCA in 2008 and a suite of documents under the general title of RICS New Rules of Measurement, which included the first edition of NRM1 New Rules of Measurement—Order of Cost Estimating and Cost Planning for Capital Building Works, in 2009.

As is the way with these things, these were followed by new editions of both these documents in 2012 [5,6].

NRM1 provides a data structure for Elements, designed Elements, sub-Elements, and components and rules of measurement; the SFCA provides detailed definitions of the Elements and rules for cost analysis. The elemental data structure is common to both documents and is referred to in this paper as the "SFCA Elements."

NON-ELEMENTAL COST CATEGORIES

Although the SFCA defines Elements as "physical parts of a building" it also includes rules for classifying costs that are normally shown separately in contract documents such as contractors, site overheads (preliminaries in UK parlance), profit, design fees, risk (contingencies), etc.

There has always been some debate about how to show these costs, but the convention is to split them out in analysis and to spread preliminaries and profit over the Elements for benchmarking purposes.

UK CLASSIFICATION WARS PART 1

The RICS was not the only organisation in the UK or elsewhere, attempting to classify Elements.

The SfB classification system ("Samarbetskommittén för Byggnadfrågor") was developed in Sweden in the late 1940s for organising specifications. It was used as the structure for the Swedish national specification in 1950. A UK version was published in 1961 [7]. It was taken up by the Conseil International du Bâtiment (CIB), now the International Council for Research and Innovation in Building and Construction, who published an international version in English and French as Building Classification Practice in 1966. This formed the basis of many classification systems around the world.

In the UK it was adopted as the basis of a national classification system—*CI/SfB Indexing Manual* [8]. This was primarily used for classifying documents in libraries, and specification information. There were attempts to use the Elements table from CI/SfB to collect and present cost information, particularly in architectural publications, but by the early 1970s the SFCA Elements had become a defacto standard.

A new classification system Unified Classification for the Construction Industry (Uniclass) [9] was developed in the UK in 1997 to replace CI/SfB. It contained a table of Elements, which, for the benefit of librarians and others, allowed the same parts of a building to be classified in different ways. It was never intended for or used for the classification of costs.

Uniclass did introduce the concepts of "Entities" (see below) and propose for the first time elements for use on civil engineering projects.

ISO 12006.2

Uniclass followed the framework set out in the ISO Technical Report 14177 Classification of Information in the Construction Industry [10], which formed the basis of ISO 12006.2 Building Construction – Organization of Information About Construction—Framework for Classification of Information [3].

During development of the ISO, attempts were made to reach agreement on international classification tables for some concepts, including Elements. These discussions foundered not only on the differing national views but also on the views of different disciplines. In the end, these efforts were abandoned and ISO 12006 defined the tables (concepts) that should be classified in a compliant national classification system.

ISO 12006.2 is currently being reviewed and a 2014 edition is out for consultation in the UK.

EUROPE—UNDERSTANDING A MISUNDERSTANDING

The European Committee for Construction Economics CEEC was set up over 20 years ago as a pan European group of professional associations in the area of construction economics. RICS represents the United Kingdom.

In the diverse European construction industry, many national industries developed the profession of construction economist, advising building and infrastructure clients or contractors on how to achieve best value for money. Where it is not recognised as a separate discipline, it has become in many cases a specialisation for architects, engineers, or project managers who have used the skills of the construction economist to give added value to their own discipline.

Two important aims of the CEEC are harmonisation of working methods and the exchange of information. *The Code of Measurement for Cost Planning* [11] was produced by a CEEC working group to enable exchange of data at high level, while still permitting differing national approaches and new innovative local developments.

The need for the code became apparent following a conference in Lausanne, Switzerland, in 1998, at which representatives from each CEEC member country were invited to describe the main method used to estimate a construction cost based on a quantitative description, and to apply that method to a common example.

The common example was a new building for a research institute of advanced technologies, to be built at the university campus of Roskilde, some 30 km west of Copenhagen. The building comprises 1800 m^2 office space. The degree of detail provided corresponds to layout drawings on scale 1:300, with an outline specification and an elemental bill of quantities. The quantities were based on the Danish elemental classification and provided a common basis for all estimates.

The objectives of the exercise were to:

- Compare approaches to cost planning
- See how Element classification varied
- Compare cost data sources
- Compare how costs were adjusted for:
  (i) Scale
  (ii) Location
  (iii) Time

The use of Elements for early cost estimates (conceptual estimates), cost planning, and cost analysis was well established in most of the countries involved as was the use of costs/m^2 of floor area for benchmarking costs.

It was accepted that, while the resultant cost plans and presentations would be indicative of the practices and procedures in each country, they would also be very much influenced by the personal practice and experience of each estimator.

The results presented and the ensuing discussion produced some interesting results:

- All countries used elemental estimating or cost planning systems
- The Elements used were similar, but were grouped and coded with differing cost classification systems
- Each country considers their system of classification and measurement as superior to the others, or at least more appropriate to their national design and procurement processes.

- The often-used comparison of square metre prices is misleading as costs included differ and the floor areas are measured differently.

It was accepted that the results of the estimates would not be directly comparable for a range of reasons:

- Quality assumed from the outline specifications
- Differences of standard specifications between countries
- Differences in national, and local, building regulations and codes
- What was included in the total costs
  (i) Taxes
  (ii) Fees
  (iii) Land costs
  (iv) Etc
  (v) The measurement of the floor area

Notwithstanding these reservations, the temptation to compare the results and to infer something about the relative costs of construction in different European countries was overwhelming. The results are shown in **Table 1**.

Indexing the results produced the graph in **Fig. 2**.

Even though there had been a long held suspicion that comparing costs between countries was dangerous, the results came as a bit of a shock, particularly as cost per m^2 tended to be the lingua franca in any discussions about costs between different countries.

The CEEC set up a committee to try and find some way of harmonising the data in order to arrive at a set of truly comparable costs in the future by bringing in representatives from various countries with clearly defined elemental classification systems to:

- Compare the lists of Elements
- Produce a grouping of Elements that could be defined from all the individual elemental systems

TABLE 1 Total cost, floor area, and cost/m^2 for research institute model.

	Total Cost ECU	Floor Area m^2	Cost Per Unit Floor Area ECU/m^2
Switzerland	2 917 000	2875	1,015
Holland	2 372 700	3007	789
France	2 138 300	3412	627
UK	2 910 370	2585	1,126
Ireland	2 900 006	2585	1,122
Finland	2 607 000	2758	945
Denmark	2 639 000	1800	1,466
Spain	1 111 650	1800	618

Note: the costs are at 1998 prices in European Currency Units (ECU) the forerunner of the Euro so that the exchange rate for all countries except UK and Switzerland were fixed.

FIG. 2 Index of total cost, floor area, and cost/m^2 (averages = 100).

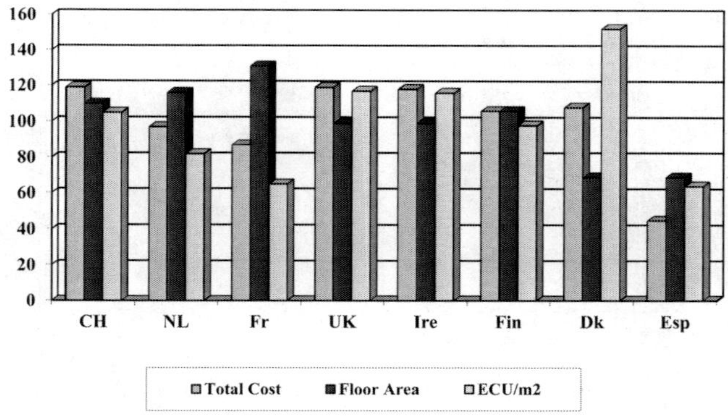

- Define any other cost groupings that might, on occasion, be included in total building costs
- Allow for the clear identification of the scope of costs included in total costs on an individual project
- Define the measurement of floor areas in such a way that the floor area used on an individual project can be identified with clarity.

Acknowledgment must be given to Martin Wright of PBK AG in Switzerland for really bringing the code together and driving the whole process through to completion.

CEEC CODE OF MEASUREMENT FOR COST PLANNING

The CEEC Code [11] provides a standard basis for the sub-division of costs and for measurement of basic quantities of buildings for international budgeting, comparison, and analysis at management level. The structure is organised to permit the use of existing national classifications at a more detailed level of information (subsidiarity principle) see Fig. 3.

The current situation can be summarised as follows:

- In European countries, different rules and definitions exist for assessing costs and quantities.
- Due to the differing national definitions, data is difficult to interpret and comparative data likely to be inaccurate.
- Comparative data is an eminently important basis for decisions by construction economists, investors, planners, administrators, and politicians.

The CEEC code aims to clarify matters by defining the typical cost classification areas used and cross-referencing to local definitions. The current Code is presented in English, French and German.

FIG. 3 Relationship of CEEC code to national classifications.

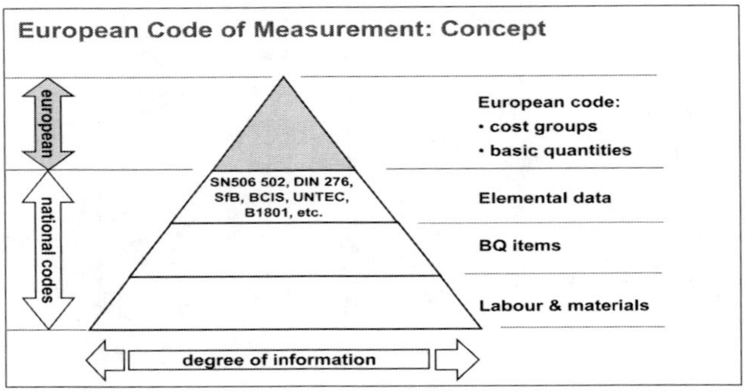

An updated 2014 edition has been agreed and is currently being translated into French and German while the participating countries complete the mapping process.

Costs

The Code defines groupings of costs that provide a framework to cover the global cost of buildings, including the costs of development, land, and life cycle costs, which are defined in some national codes. This goes further than traditional practice in some countries and groups costs into five blocks:

- Construction costs and fees
- Incidental costs
- Costs in use
- Site acquisition
- Project funding

This permits overall project appraisal, and if items are not included in individual countries, this will be clearly apparent and will avoid misunderstandings on the overall scope of the costs.

Areas

Definitions of quantities have been restricted to 12 basic quantities for site areas, floor areas, and functional units. Elemental quantities have not been defined, as local definitions may be more suitable for analysis of elemental unit rates.

The list of cost groups is given in Appendix A.

A mapping of UNIFORMAT II, ASTM E1557-09 [12], to the Code has been included in Appendix B.

WIDER INTERNATIONAL COMPARISONS

BCIS has had a wide dialogue with like-minded organizations around the world, including, from the inception of UNIFORMAT II, ASTM. This has always been a valued and fruitful relationship and being involved in the ASTM discussion groups on the developing civil engineering Elements tables has been a huge help in clarifying our own thoughts.

In 2009, BCIS carried out a survey of the use of elemental classification for costing among RICS members around the world [13]. Because of the source of the responses, BCIS does not claim this survey to be exhaustive.

It presents information on practices in 40 different countries gleaned from 103 survey responses. The relevant findings were:

- 19 out of the 40 countries reported on did not have a published standard elemental classification of building parts for costing purposes.
- 40 standards were referenced although the references where not always precise enough to track down, and some of the standards identified were clearly not "elemental" as BCIS, or ASTM would understand the term.
- Many countries where such standards exist tended to be linked historically with the United Kingdom.
- In the absence of locally agreed standards, professionals commonly adopt "foreign" standards or ad hoc in-house developed standards.
- The use of ASTM UNIFORMAT II was referenced in the United States, Canada, and United Arab Emirates.
- The use of the SFCA was referenced in the UK, Australia, Barbados, Cyprus, Malaysia, Mauritius, Nigeria, St Kitts and Nevis, Trinidad and Tobago, and United Arab Emirates.

In response to the survey findings, BCIS published Principles of elemental classification for buildings (International) [14] to promote the use of Elements generally and to offer the SFCA Elements as the starting point for a standard where none exists. The writers do not know if anyone has taken this up. It is currently available in English, Chinese, and Spanish.

Why Entities?

What works for buildings should work for civil engineering. The ISO committee looking at Elements as part of the development of ISO 12006.2 [3] spent a lot of time pondering why the use of Elements had not been applied to civil engineering. The penny dropped when someone asked what we meant by "building" and what we meant by "civil engineering" and put the following on a flip chart (see **Table 2**).

TABLE 2 Comparisons of building and civil engineering attributes.

Building is:	Civil Engineering is:
A process—building	A process—civil engineering
An industry—building	An industry—civil engineering
A product—a building	A product—???

It is the product “a building” that can be defined by its “object function” and broken down into Elements. When we began to search for types of civil engineering structures that have a common “object function,” we were producing a new classification table which we called “Entities.”

Entities are defined in Uniclass as “independent construction of significant scale classified according to their physical form/basic function.”

Having classified the Entities, the Elements were established for each Entity and the definition of an Element adjusted accordingly: “a major physical part or system of a building or other construction Entity, which, in itself or in combination with other parts or systems, fulfils a characteristic predominating function of the building or other construction Entity, considered without regard to the type of technical solution or method of form of construction.”

The requirement for an Entities table was included in ISO 12006.2 [3], where it is defined as “independent material construction result of significant scale serving at least one user activity or function.”

As part of research into the use of Elements for civil engineering, BCIS carried out an Internet/desk research to identify any classification of Entities. It unearthed 14 sources that contained, at least partial, lists of Entities. They are listed in Appendix C.

ELEMENTS FOR CIVIL ENGINEERING STRUCTURES

BCIS has hosted building cost analyses for the UK Defence Infrastructure Organisation (DIO) for many years. In 2009, DIO asked BCIS to develop a standard form of cost analysis and set up a database for their civil engineering projects.

The resulting Standard Form of Civil Engineering Cost Analysis [15] classifies Entities and provides a definition. For Example:

Pavements

Definition: Unenclosed usable surfaces

Includes: Roads, runways, hard landscaping, etc.

Primary functional unit: Surface area fulfilling primary function (m2)

Railways and the like

Definition: Railways on which trains or similar transport run

Includes: Railways, tramways, etc.

Primary functional unit: Equivalent track length of track (km)

Quays, Piers, Jetties, Platforms and the like

Definition: Usable surface with provision for mooring, embarkation, etc. to one or more sides

Includes: Quays (marginal berths), jetties, piers, platforms, etc.

A full list is included in Appendix D.

BCIS has developed an outline elemental data structure for each entity and detailed definition for those Entities where we are collecting data for DIO.

In drafting the definitions, it became apparent that, while an Entity always has the same Elements they needed to be defined in terms that are relevant to different user activities. For example, the Elements for pavements are:

Pavements

1 Substructure

- 1A Subgrade

2 Pavements

- 2A Preparation (for resurfacing)
- 2B Pavement structure
- 2C Pavement surface
- 2E Edge treatment

3 Pavement completion

- 3A Drainage
- 3B Electrical power installation
- 3C Lighting installation
- 3D Heating installation
- 3F Communications
- 3G Protection and security
- 3H Special installation
- 3I Ancillary (structures and equipment)
- 3J Builders Work In Connection

The writers drafted some generic definitions but needed to draft separate special definitions for runways and roads to reflect the terminology and which were acceptable to specialists.

For example, the "Communications" Element includes signs, signals, and guidance lighting. It was difficult to persuade the runway specialists that traffic lights were the same Element as Air Guidance Lighting.

This data structure has been implemented for the DIO and issued as a consultation document to the rest of the industry as the BCIS Standard Form of Civil Engineering Cost Analysis.

The consultation documents consist of the following:

- Outline data structure: key consultation document that describes the general concepts and shows the proposed data structure.
- Principles, instructions, definitions, and common Elements: Sets out the principles of analysis; details the supporting information required so that the costs analysed can be fully understood; sets out the cost breakdown for works that occur on most projects (external works, facilitating works) and for preliminaries, temporary works, design fees, etc.

Definitions of Elements for some individual entities: provides classification of the entity types and detailed definitions of the Elements:

- Pavements, which cover roads, runways, hard standings, etc. Includes specific definitions for Runway pavements.
- Railways and the like

- Quays, Piers, Jetties, Platforms and the like
- Support structures (Masts, Towers and the like)
- Pipelines and Ducts

A link to these documents is included in the Ref. [15].

UK GOVERNMENT CONSTRUCTION STRATEGY

In 2011, the UK government launched its construction strategy [16], which called for "a profound change in the relationship between public authorities and the construction industry to ensure the government consistently gets a good deal and the country gets the social and economic infrastructure it needs for the long-term." The government set itself the target of reducing costs by 20 % over the next five years.

The three key aspects of this strategy were procurement, benchmarking, and building information modelling (BIM).

The objective of the benchmarking is to set a challenging market price for procurement of the asset, so that cost becomes a major driver in project design and delivery.

Government departments are required to develop "Should cost" capability, i.e., where an intelligent client understands what a particular requirement should cost before going to market. Typically, this knowledge is gained by reference to earlier cost benchmarks, together with an appreciation of what is currently affordable, e.g., those costs towards the bottom of the cost distribution for a particular type of project. It is therefore envisaged that "should cost" capability would be deployed together with output/outcome specifications in order to ensure the final specification represents the minimum needed to effectively deliver a client's business requirement.

There is a stated objective for the development of protocols to capitalise on BIM in the collection and use of cost data.

Most departments and agencies had benchmark data which they already used to assess value for money on individual projects. These consist of cost data and ancillary data to give context to individual projects and to identify cost drivers. These have been used as the basis for a consistent set of cost benchmarks that are published through the Cabinet Office Data and Benchmarking Task Group [17], which has defined four types of benchmark:

- Type 1 benchmark; Spatial measures—£/m^2, £/m, £/km, £/m^3
- Type 2 benchmark: functional measures—business outcomes £/place in schools, £of flood damage avoided/£invested
- Type 3 benchmark: department specific—e.g., MoJ ratio of production cost/total construction cost
- Type 4 benchmark: type 1 at the elemental level

The Departments and Organisations providing benchmarks through the Cabinet Office Data and Benchmarking Task Group are:

- Department of Health (DoH/P21)
- Environment Agency (DEFRA/EA)

- Education Funding Agency (DfE/EFA)
- Highways Agency (DfT/HA)
- Homes and Communities Agency (DCLG/HCA)
- Ministry of Defence (MoD)
- Ministry of Justice (MoJ)
- Network Rail (NR)
- London Underground (LU)

Each department was required to set a cost reduction trajectory to indicate how it would achieve the required savings.

For buildings, all departments used the SFCA Elements for collecting costs; the scope of the individual department benchmarks are defined by reference to it.

The use of elemental benchmarks has allowed for some meaningful comparisons between departments.

Because the department and agencies delivering civil engineering infrastructure are, in the main, constructing different types of structures (Entities and Entity types), they have developed individual cost breakdown structures. These are broadly elementally based, but have been developed for specific users and often contain a mixture of Entities, Entity types, Elements, Components, work results, and type of work.

As part of the work BCIS is doing with the Benchmarking Task Group, we carried out an analysis of the data used by departments for benchmarking and identified the following classification tables:

- **Entity by function and form:** independent construction of significant scale classified accordance with its basic function (i.e., spanning—e.g., Bridge) and form (e.g., simple span bridge).
- **Entity types defined by user activity or function:** entity classified by what it is used for, e.g., rail bridge, acute hospital, court building.
- **Costs classified by:**
  - o Elements

    Major physical part of an entity that fulfils a specific function irrespective of its design specification or construction, e.g., external wall to a building, abutment to a bridge.
  - o Non-Elemental cost categories.

    Costs incurred in delivering the entity, e.g., preliminaries, temporary works, fees.
- **Type of work:** the work carried out to the entity or Element—e.g., new build, refurbishment, track renewal, re-ballasting.
- **Spaces by usability for buildings:** e.g., circulation, ancillary, internal divisions, usable.
- **"Spaces" by user activity:** ("spaces" = subdivision of entity) space defined by what it is used for, e.g., general teaching and ICT, learning resource areas, ward accommodation.
- **Cost influencing factors:** classification of circumstances that influence costs—e.g., site conditions—contour, site conditions—ground conditions, listed building.

THE COMING OF BIM

BCIS has been engaged with the concepts that have become BIM from the 1990s, through involvement with the International Alliance for Interoperability (IAI) and BuildingSmart because the transmission of data between systems requires a data structure and classification.

The government's Construction Strategy requires that by 2016, all government projects should be delivered using a BIM [18].

Rather than define BIM, the strategy sets out a hypothesis that "Government as a client can derive significant improvements in cost value and carbon performance through the use of open shareable asset information."

The requirement is for, what it defines as Level 2 BIM, i.e., "Managed 3D environment held in separate discipline 'BIM' tools with attached data. Commercial data managed by ERP (Enterprise Resource Planning) software integration on the basis of proprietary interfaces or bespoke middleware. The approach may utilise 4D programme data and 5D cost Elements as well as feed operational systems."

The ultimate aim is for, what it defines as, Level 3 BIM, i.e., "a fully open process and data integration enabled by "web services" compliant with the emerging IFC/IFD standards, managed by a collaborative model server. Could be regarded as iBIM or integrated BIM potentially employing concurrent engineering processes".

The means of data exchange at Level 2 is mandated as the use of Construction Operations Building information exchange (COBie)[3], a non-proprietary format based on a spreadsheet so it can be managed by organisations of any size at any level of IT capability but can be linked to other systems and software.

For Level 3, the data exchange is likely to be via the Industry Foundation Classes (IFC)[4].

To use COBie and IFCs to exchange data consistently requires a common classification.

UK CLASSIFICATION WARS PART 2

The maintenance of Uniclass had become the responsibility of a joint industry body, the Construction Industry Project Information Committee (CPIC). The uptake of the tables for layering conventions and object classification in CAD modeling had confused their use and there have been regular calls for them to be completely updated.

The CAD modelers had latched onto the Elements table in Uniclass and taken it places that it was not intended to go. In the absence of an appropriate table to classify design solutions, they had ignored the "without regard to the type of

---

[3]COBie is an information exchange specification for the life-cycle capture and delivery of information needed by facility managers developed by the National Institute of Building Science (see http://www.nibs.org/?page=bsa_cobie).

[4]Industry Foundation Classes a common data schema for exchanging data between software applications, developed by BuildingSmart (see http://www.buildingsmart.org/standards).

technical solution or method of form of construction" in the definition of the table and extended it to include design solutions and specification.

The original Uniclass was funded by three professional institutions, a contractors' representative body and the government, but much of the work was done by volunteers from the funding organizations and other industry bodies.

There was no funding available to manage a review and updating, and while CPIC tried to organize volunteer groups, little progress was made. Some funding was allocated by one participating organization, but the proposals that came forward were produced with no consultation and no regard for legacy systems or the classifications in use within the industry and; as a result, have found little favour.

Thus Uniclass had rather lost its way; some tables had been updated, some alternatives suggested, some tables abused, and many simply ignored. It did not meet the requirements of a classification system for use in a BIM and the growing importance of BIM increased the demand for a new classification system.

Because one of the most widely used BIM tools, Autodesk's Revit, has OmniClass[5] embedded in the software, it has been suggested as an alternative to Uniclass.

BIM tools developed for different disciplines used the classification appropriate to that discipline; as such, most costing tools have implemented the SFCA Elements and the NRM data structures. Those that have originated in other countries will also have data structures that reflect the cost databases that they reference; for US based systems, this is very often R.S. Means estimating data arranged in the CSI Masterformat.2010 system[6].

UK GOVERNMENT DIGITAL TOOL FOR BIM

In order to address the need for Classification and some other issues around the delivery of its BIM target, the government's BIM Task Group have initiated a competition to deliver a "digital tool for information modelling" that will define geometry, scope of service, and data deliveries across all disciplines at various key stages of a project life at a level of detail that is significant enough to work, but no more.

One of the key outputs of this initiative will be a new classification that embraces legacy systems, is future proof and will integrate the design, delivery, and operation information requirements.

REVISION TO ISO 12006.2

The ISO on organization of information about construction works is being revised and a draft is currently (March 2014) out for comment.

One of the UK government's requirements for a new classification is that it should comply with the new edition of the ISO.

[5] The OmniClass Construction Classification System (known as OmniClass or OCCS) is designed to provide a standardized basis for classifying information created and used by the North American architectural, engineering and construction (AEC) industry (see http://www.omniclass.org/index.asp).

[6] CSI Masterformat is a master list of numbers and titles classified by work results published by the Construction Specifications Institute (CSI) (see http://www.csinet.org/).

The good news is that the draft contains a requirement for both "Construction Entities" and "Construction Elements" tables and that there is a note on the definition of an "Element" to the effect that "for practical purposes, such as carrying out cost analysis of a construction Entity, it is vital that construction Elements are mutually exclusive, in order to ensure that each part is counted only once."

The bad news is that the definitions have been changed in such a way that they blur their meaning, or perhaps they are intended to change the meaning.

Element:

"Construction Entity part which, in itself or in combination with other such parts, fulfils a predominating function of the construction Entity." ISO 12006.2 2001.

"Constituent of a construction Entity with a characteristic technical function, form or position." ISO 12006.2 2014 (Draft dated 16 January 2014).

INTERNATIONAL PROPERTY MEASUREMENT STANDARD

RICS and ASTM are members of the International Property Measurement Standard (IPMS) Coalition[7], the aim of which is to enable properties to be measured on a transparent basis that promotes market efficiency through greater confidence between investors, occupiers, and funds. The growth of cross-border property investment and expansion by global corporate occupiers underpins the demand for transparency against the background of many differing national and local measurement conventions.

IPMS have been developed to work with associated financial reporting and valuation standards such as International Financial Reporting Standards (IFRS) and the Uniform Standards of Professional Appraisal Practice (USPAP) in the USA.

While IPMS is focused on measurement for property (real estate) purposes there is a desire to extend this to construction, and discussions have started with CEEC, and other members of the IPMS coalition with an interest in construction, to see if their Code can act as a template to define the scope of construction work and measurement in any international cost per unit comparisons of construction costs.

Conclusion

The construction industry has come a long way since the 1950s and the need for classification has expanded with the introduction of computers and BIM. As a result, there are now many more parties with an interest in classification. However, the use of Elements to model costs from the earliest stage of a project is still the chief concern of the cost consultant.

It is therefore important that the concept of Elements defined as "a major physical part or system of a building or other construction Entity, which, in itself or in combination with other Elements, fulfils a characteristic predominating function of the building or other construction Entity, considered without regard to the type of

[7] International Property Management Standard Coalition (see http://ipmsc.org/).

technical solution or method of form of construction" is defended and embodied in wider classification systems.

As the world shrinks, the demand for internationally consistent classification of construction costs will grow. In the first instance this will result in a lot of mapping exercises between national systems and the development of concordance documents like the CEEC code. This may be enough, but now might be the time to work with others to explore the holy grail of a single elemental cost breakdown structure.

Appendix A

TABLE 3 CEEC cost groups.

CONSTRUCTION COSTS and FEES	
#01	Substructure
#02	External superstructure/envelope
#03	Internal superstructure
#04	Internal finishings
#05	Services installations
#06	Special equipment and installations
#07	Furniture and fittings
#08	Prefabricated buildings, building units and pods
#09	Site and external works
#10	Site preparation
#11	Construction site overheads and management (Preliminaries)
#12	Design and project team fees
#13	Taxes on construction costs and Fees
INCIDENTAL COSTS	
#14	Ancillary costs and charges
#15	Project budget risk allowances (contingencies)
#16	Taxes on incidental costs
COSTS IN USE	
#17	Maintenance
#18	Operation
#19	End of life
#20	Taxes on cost in use
SITE ACQUISITION	
#21	Site acquisition costs
#22	Taxes on site acquisition
PROJECT FUNDING	
#23	Finance
#24	Grants and subsidies
#25	Taxes on project funding

Note: Source, CEEC Code of Measurement Practice.

Appendix B

TABLE 4 Mapping UNIFORMAT II (ASTM E1557) to the CEEC code of measurement for cost planning.

CEEC Code of Measurement for Cost Planning

UNIFORMAT II (ASTM E1557-09)

Mapping of cost groups. Note: I have only mapped E1557 to the CEEC construction elements #1 to #10. Other CEEC cost categories are covered in ASTM E2083-05 Standard Classification for Building Construction Field Requirements and Office Overheads Profit [19]; ASTM E2168-10 Standard Classification for Allowance, Contingency and Reserve Sums in Building Construction Estimating [20]; and ASTM E2620-08 Standard Classification for Program and Project Estimate Summaries [21].

Principles for cost information

Cost information should always include the base date, exchange rates and, in the case of costs in use, the period in time being evaluated. In addition the project time scale for construction including planning and approval should be stated.

Costs will generally be placed in the most suitable category; minor deviations due to restrictions of national coding should be noted.

In some countries it may be difficult to subdivide costs into the cost groups. In these cases groups may be combined for analysis purposes.

The limits of any costs (e.g., costs within site boundaries) should be clearly stated.

CONSTRUCTION COSTS and FEES

#01	**Substructure**	
	UNIFORMAT II (ASTM E1557-09)	
	A10	Foundations
	A2010	Basement Excavation
#02	**External superstructure/envelope**	
	UNIFORMAT II (ASTM E1557-09)	
	A 2020	Basement Walls
	B20	Exterior Enclosure
	B30	Roof
#03	**Internal superstructure**	
	UNIFORMAT II (ASTM E1557-09)	
	B10	Superstructure
	C10	Interior Construction
	C20	Stairs
#04	**Internal finishings**	
	UNIFORMAT II (ASTM E1557-09)	
	C30	Interior finishes
#05	**Services installations**	
	UNIFORMAT II (ASTM E1557-09)	
	D	Services
#06	**Special equipment and installations**	
	UNIFORMAT II (ASTM E1557-09)	
	E10	Equipment
	F1030	Special Construction Systems

TABLE 4 *(Continued)*

	F1040	Special Features
	F1050	Special Controls and Instrumentation
#07	**Furniture and fittings**	
	UNIFORMAT II (ASTM E1557-09)	
	E20	Furnishings
#08	**Prefabricated buildings, building units and pods**	
	UNIFORMAT II (ASTM E1557-09)	
	F1010	Special Structures
	F1020	Integrated Construction
#09	**Site and external works**	
	UNIFORMAT II (ASTM E1557-09)	
	G20	Site Improvements
#10	**Site preparation**	
	UNIFORMAT II (ASTM E1557-09)	
	F20	Selective demolition
	G10	Site preparation
#11	**Construction site overheads and management**	
	UNIFORMAT II (ASTM E1557-09)	
		Not covered
	Standard Classification for Building Construction Field Requirements, and Office Overheads & Profit (ASTM E2083-05 (2010))	
	X10	Field Requirements
	X20	Office Overheads and Profit
#12	**Design and project team fees**	
	UNIFORMAT II (ASTM E1557-09)	
		Not covered
#13	**Taxes on construction costs and fees**	
	UNIFORMAT II (ASTM E1557-09)	
		Not covered
INCIDENTAL COSTS		
#14	**Ancillary costs and charges**	
	UNIFORMAT II (ASTM E1557-09)	
		Not covered
#15	**Project budget risk allowances (Contingencies)**	
	UNIFORMAT II (ASTM E1557-09)	
		Not covered
	Standard Classification for Allowance, Contingency and Reserve Sums in Building Construction Estimating (ASTM E2168-10)	
#16	**Taxes on incidental costs**	
	UNIFORMAT II (ASTM E1557-09)	
		Not covered

TABLE 4 *(Continued)*

COSTS IN USE		
#17	**Maintenance**	
	UNIFORMAT II (ASTM E1557-09)	
		Not covered
#18	**Operation**	
	UNIFORMAT II (ASTM E1557-09)*SMLCC*	
		Not covered
#19	**End of life**	
	UNIFORMAT II (ASTM E1557-09)*SMLCC*	
		Not covered
#20	**Taxes on cost in use**	
	UNIFORMAT II (ASTM E1557-09) *HM Revenue and Customs*	
		Not covered
SITE ACQUISITION		
#21	**Site acquisition costs**	
	UNIFORMAT II (ASTM E1557-09)	
		Not covered
#22	**Taxes on site acquisition**	
	UNIFORMAT II (ASTM E1557-09)	
		Not covered
PROJECT FUNDING		
#23	**Finance**	
	UNIFORMAT II (ASTM E1557-09)	
		Not covered
#24	**Grants and subsidies**	
	UNIFORMAT II (ASTM E1557-09)	
		Not covered
#25	Taxes on project funding	
	UNIFORMAT II (ASTM E1557-09)	
		Not covered
COSTS INCLUDED IN CEEC CODE NOT INCLUDED IN UNIFORMAT II STANDARD		
#11	Construction site overheads and management	
#12	Design and project team fees	
#14–#15	Incidental costs	
#17–#19	Costs-in-use	
#21	Site acquisition	
#23–24	Project funding	
#13, #16, #20, #22, #25	Taxes	

Appendix C

TABLE 5 Classification of entities.

Country	Classification Name	Entities
Australia	NATSPEC	Bridges, Roadways, Public Utilities
Czech Republic	URS Praha Specification System	General structures and work, Vertical and complex structures, Horizontal structures, Roads and railroads, Structures and associated construction production works, Pipe lines, Other structures and work, demolitions
Germany	DIN 276	Traffic installations, Roads, Track systems, Protective structures, Bridges, Shaft systems, Hydraulic systems, Water supply systems
Hong Kong	General Specification For Civil Engineering Works Volumes 1, 2 & 3	Carriageways, Bridges, Marine works, Railway works and Utilities
Netherlands	RAW	Piping, Cabling, Road Pavements, Rail and Tram Works
South Africa	SANS 1200	Road and Railway
Sweden	BSAB 1996 (Draft)	Load carrying structures, Pavements, Pipe and Ductwork, Electrical and Telecommunications Systems, Transport
United Kingdom	Uniclass (Table H)	Pavement, Tunnels, Shafts, Cuttings; Embankments, Retaining walls, Tanks, Bridges, Pipelines, Towers
	BCIS Civil Engineering Cost Analysis	Transport, Public Health, Service Industries, Process Industries, Environmental
	Defence Estates	Roadworks, Earthmoving, Drainage Sewage Treatment, Bridgeworks, Small arms ranges, Masts and towers, Airfield pavement works
USA	OmniClass	Buildings, Non Building Structures, Movable Structures, Land Forms, Water Forms
	Masterformat	Utilities, Transport and Marines
	ASTM: summary of Element lists	Entities for which there are standards or draft standards include: Buildings, Bridges, Highways, Dams & canals, Waterfront, Road tunnels.
International ISO 12006.2	Construction Entities	List of example

Source: BCIS survey.

Appendix D

TABLE 6 BCIS SFCECA entities table.

Buildings
Definition: Enclosed usable floor space
Includes: Buildings
Primary functional unit: Gross internal floor area (m^2)
Pavements
Definition: Unenclosed usable surfaces
Includes: Roads, runways, hard landscaping, etc.
Primary functional unit: Surface area fulfilling primary function (m^2)
Landscaping
Definition: Unenclosed planted surfaces
Includes: Soft landscaping
Primary functional unit: Surface area fulfilling primary function (m^2)
Railways and the like
Definition: Railways on which trains or similar transport run
Includes: Railways, tramways, etc.
Primary functional unit: Equivalent track length of track (Km)
Quays, Piers, Jetties, Platforms and the like
Definition: Usable surface with provision for mooring, embarkation, etc. to one or more sides
Includes: Quays (marginal berths), jetties, piers, platforms, etc.
Primary functional unit: Length of mooring, embarkation, etc provided (m)
Retaining structures
Definition: Vertical retention of solids, powders and liquids
Includes: Embankments, retaining walls, sea walls, riparian walls, etc.
Excludes: Embankments formed of the retained medium, see *Ground contouring*
Primary functional unit: Area of retained substance measured on a vertical plane (m^2)
Dividing structures
Definition: Vertical division of horizontal surfaces
Includes: Walls, fences, etc.
Primary functional unit: Area measured on a vertical plane (m^2)
Containing structures
Definition: Containment of solids, powders, liquids and gases
Includes: Tanks, silos, etc.
Primary functional unit: Volume of contained substance (m^3)
Support structures (Masts, Towers and the like)
Definition: Supporting structures
Includes: Masts, towers, gantries, windmills, etc.
Primary functional unit: Height (m)

TABLE 6 *(Continued)*

Pipelines and Ducts (underground)
Definition: Enclosed conduits for the flow and containment of solids, powders, liquids, and gases
Includes: Pipes, ducts
Primary functional unit: Length (m)
Pipelines and Ducts (supported)
Definition: Enclosed conduits for the flow and containment of solids, powders, liquids, and gases
Includes: Pipes, ducts, etc.
Primary functional unit: Length (m)
Open conduits
Definition: Open conduits for the flow of solids, powders and liquids
Includes: Canals, channels, ditches, etc.
Primary functional unit: length (m)
Cables (buried)
Definition: Conductors for electricity and telecommunications
Includes: Cables, power lines, etc.
Primary functional unit: Length (m)
Cables (suspended)
Definition: Conductors for electricity and telecommunications
Includes: Cables, power lines, etc.
Primary functional unit: Length (m)
Wells
Definition: Shafts for collection of liquids
Includes: Wells, boreholes, etc.
Primary functional unit: depth (m)
Ground contouring
Definition: Reformation of the surface of the earth.
Includes: Earthworks, ground remediation, earth embankments, dredging, etc.
Primary functional unit: Volume of treated earth (m3)
Bridges (spanning structure)
Definition: Spanning structures
Includes: Bridges, viaducts, etc.
Primary functional unit: Length of span (m)
Tunnels
Definition: Underground enclosed connecting structures
Includes: Tunnels
Primary functional unit: Length (m)

References

[1] Ministry of Education, *2Building Bulletin No 4—Cost Study2, HMSO*, Ministry of Education, London, 1951.

[2] BCIS, *Standard Form of Cost Analysis: Principles, Instructions and Definitions*, Royal Institution of Chartered Surveyors, London, 1969.

[3] ISO 12006.2 (2001): Building Construction - Organization of Information About Construction Works—Part 2: Framework for Classification of Information, ISO, Geneva, Switzerland, 2001.

[4] RICS, *SMM7—Standard Method of Measurement for Building Works*, 7th ed., Royal Institution of Chartered Surveyors, London, 1998.

[5] RICS, *NRM 1, RICS New Rules of Measurement; Order of Cost Estimating and Cost Planning for Capital Building Works*, London, 2012.

[6] BCIS, *Elemental Standard Form of Cost Analysis: Principles, Instructions Elements and Definitions*, 4th (NRM) ed., Royal Institution of Chartered Surveyors, London, 2012.

[7] RIBA, *SfB/UDC Building Filing Manual: Recommendations for Standard Practice in Preclassification and Filing*, RIBA, London, 1961.

[8] CIB, "Building Classification Practice," *CIB Report No. 6*, CIB, Rottendam, The Netherlands, 1966.

[9] CC, RIBA, RICS, CIBSE, DoE, *Uniclass—Unified Classification for the Construction Industry*, RIBA, London, 1997.

[10] ISO, "Classification of Information in the Construction Industry," *Technical Report 14177*, ISO, Geneva, Switzerland, 1994.

[11] CEEC, 2003, "CEEC Code of Measurement for Cost Planning," http://www.ceecorg.eu/eurupean-code (Last accessed 17 Oct 2012).

[12] ASTM E1557-09: Standard Classification for Building Elements and Related Sitework—UNIFORMAT II, *Annual Book of ASTM Standards*, ASTM International, West Conshohocken, PA, 2009.

[13] BCIS, 2009, "2009 BCIS International Cost Elements Enquiry Report," Royal Institution of Chartered Surveyors, London, http://www.rics.org/Global/2009-BCIS-IntCostEnquiry Rpt.pdf (Last accessed 13 Nov 2013).

[14] BCIS, 2012, "Principles of Elemental Classification for Buildings (International)," Royal Institution of Chartered Surveyors, London, http://www.rics.org/uk/knowledge/bcis/about-bcis/forms-and-documents/bcis-principles-of-elemental-classification-for-buildings-international/ (Last accessed 13 Nov 2013).

[15] BCIS, 2012, "Standard Forms of Civil Engineering Cost Analysis, BCIS, RICS," http://www.rics.org/uk/knowledge/bcis/about-bcis/forms-and-documents/standard-form-of-civil-engineering-cost-analysis/(Last accessed 13 Nov 2013).

[16] Cabinet Office, 2011, "Government Construction Strategy," https://www.gov.uk/government/uploads/system/uploads/attachment_data/file/61152/Government-Construction-Strategy_0.pdf (Last accessed 13 Nov 2013).

[17] Cabinet Office, 2013, "Construction Strategy: Construction Cost Reductions, Cost Benchmarks, & Cost Reduction Trajectories to March 2013: Addendum Dec 2013," https://www.gov.uk/government/uploads/system/uploads/attachment_data/file/266816/Annex_A_Departmental_Cost_Benchmarks_Cost_Reduction_Trajectories_and_Indicative_Cost_Reductions_Addendum.pdf (Last accessed 13 Nov 2013).

[18] UK Construction Client Group, 2011, "BIM Management for Value, Cost & Carbon Improvement: A Report for the Government Construction Client Group - Building Information Modelling (BIM) Working Party Strategy Paper," http://www.bimtaskgroup.org/wp-content/uploads/2012/03/BIS-BIM-strategy-Report.pdf (Last accessed 13 Nov 2013).

[19] ASTM E2083-05: Standard Classification for Building Construction Field Requirements and Office Overheads Profit, *Annual Book of ASTM Standards*, ASTM International, West Conshohocken, PA, 2010.

[20] ASTM E2168-10: Standard Classification for Allowance, Contingency and Reserve Sums in Building Construction Estimating, *Annual Book of ASTM Standards*, ASTM International, West Conshohocken, PA, 2010.

[21] ASTM E2620-08: Standard Classification for Program and Project Estimate Summaries, *Annual Book of ASTM Standards*, ASTM International, West Conshohocken, PA, 2008.

STP 1586, 2014 / available online at www.astm.org / doi: 10.1520/STP158620140037

Anthony L. Huxley[1]

The Philosophy and Logic within UNIFORMAT II Classifications

Reference

Huxley, Anthony L., "The Philosophy and Logic within UNIFORMAT II Classifications," *Ideas to Impact: How Building Economic Standards Keep You on Track*, STP 1586, Robert E. Chapman and Michael N. Goodkind, Eds., pp. 80–101, doi:10.1520/STP158620140037, ASTM International, West Conshohocken, PA 2014.[2]

ABSTRACT

UNIFORMAT II is a mystery to some, and its presumed simplicity has been misunderstood by others. This paper makes clear the underlying principles that guided the original (ASTM E1557), and continue to guide the development of additional, ASTM UNIFORMAT II elemental classifications. This elemental view of construction has its origins in an observed need for cost management during the planning and design phases of buildings, and it has been found to be valuable for other purposes as well. Extension of this concept into other forms of construction is following the same principles of functional classification. UNIFORMAT II classifications of construction work are unique. They do not apply the traditional trade/material classification methodology so well understood by the North American construction industry. Instead they identify, summarize, and allow control of specific, critical, and relevant functional parts, or components, without regard to any proposed design solution within an overall construction entity. In order for this to be successful, a succinct, flexible, and yet structured schema is necessary, which this paper describes. Most industry professionals specialize in their chosen industry sector. In following that general principle, the writing of discrete UNIFORMAT II classifications, tailored to entities within these sectors, is also seen as relevant and valuable. Drawn together as modules, they form an overall project, while still affording description and control at a relevant and meaningful level.

Manuscript received April 2, 2014; accepted for publication June 24, 2014; published online August 4, 2014.

[1]M.Sc., MCIOB, PQS(F), President, A. Laurence Huxley Consulting Services Limited, ON, Canada,
e-mail: info@uniformat.com

[2]ASTM Symposium on *Ideas to Impact: How Building Economic Standards Keep You on Track* on October 5, 2014 in New Orleans, LA.

Keywords
UNIFORMAT II, elemental, function, classification, construction, cost, management, control

Introduction

A FRAMEWORK

ASTM International's UNIFORMAT II construction element classification is now a framework—a logical structure established to organize construction element classifications into categories and element groupings that make it easy for users to find appropriate classifications and better understand their content.

Originally a single standard classification for buildings and related sitework alone [1], this framework now applies the same principles and includes a wider range of construction industry entities. However, few projects and fewer practitioners need a complete, all-encompassing, single package, so the framework is being developed in discrete, modular packages. This modular approach engenders the use of elemental classifications in project collections that are lightweight, independent, and feature specific. Each separate, discrete module can be used repeatedly, as often as needed, and also can be maintained and evolve more easily than if it were part of one large, single package.

Elemental classifications are structured with each individual classification following a common, fixed, internal hierarchical arrangement. Each classification permits a granular view of each entity, the sum of which may be gathered into a broader hierarchical aggregation to suit higher level reporting and control needs.

VIEWING CONSTRUCTION

Construction entities can be viewed in many ways, and depending on one's needs, one may choose from among these "views" and use them at different stages of design development.

For example, construction may be viewed

- contractually—as a series of construction contracts,
- visually—as pictures, renderings, or drawings,
- financially—as expenditures and return on investment,
- textually—as a written description,
- functionally—as parts that perform specific functions,
- temporally—as operations that take time to complete, or
- spatially—as areas for specific use.

Within each view there will likely be a need to record the various parts in a logical and repeatable sequence. These parts might be ordered by

- contract,
- function,
- material,
- time,
- location,

or any combination of these and many other designations.

In all cases the view will become a model of the completed object expressed in the selected terms appropriate to the viewer. A financial specialist will see a picture within the provided cost figures, and each specialist will see and picture that same object in terms of his or her own discipline and its end needs.

For control purposes, a perfectly suitable sequence should be "seeable" at all stages in the constantly evolving planning, design, construction, operation, maintenance, and disposal phases of any object. A common classification hierarchy is necessary in order for logical and meaningful comparisons to be made at any stage. This aids in the identification of change, and in locating the source(s) of change or difference, in an environment of ever expanding knowledge and increasing detail. In essence, the need is to create simplicity out of complexity in a model that is representative of the finished product.

Achieving a balanced and usable ideal requires a pragmatic choice of component/parts/functions that can be applied universally and meaningfully across a range of demonstrably similar structures.

UNIFORMAT II ORIGINS

UNIFORMAT II was originally developed for and used solely for buildings. This 1990s element classification [1] was a development of Uniformat developed for the General Services Administration and the American Institute of Architects in the 1970s [2]. This was itself a North American development of a concept first developed within the United Kingdom's Department of Education during the 1950s. That UK initiative went on to become the *Standard Form of Cost Analysis* [3] promoted by the Royal Institution of Chartered Surveyors, which, with occasional minor local variation, is in wide use today throughout the world (an update of that seminal work was published in February 2009 [4]). It is important to note that the format was originally drafted to control the problem of construction end cost escalation during the design process, and thus reduce the instances of construction bids exceeding initial, approved, budgets. All versions apply the principle of monitoring and controlling elements (functional elements) within the proposed building design.

In its development for elemental cost analysis, UNIFORMAT II follows specific principles that also lend themselves to a variety of other reporting and control uses. Cost management is the principal goal. Cost planning, cost control, and cost analysis are the three legs supporting that cost management stool, yet the elements used have been found to be equally suitable for descriptive and quantitative purposes.

As defined by ASTM International (ASTM E833 [5]),

> **UNIFORMAT II (UII),** *n*—a hierarchical breakdown structure of construction work ordered by elements.
>
> DISCUSSION—Primarily designed for cost management (planning, control, and analysis) during the planning, budgeting, and design phases of

construction, its hierarchical elemental breakdown structure is also used for qualitative (text-rich) reports (preliminary project description, condition assessment, asset description) and other quantitative (text and numerical) purposes (value engineering, risk analysis, preliminary time schedule, building information modeling, etc.).

UNIFORMAT II, as related to buildings and related sitework, has a significant history, a proven worth, and much commonality with similar systems in everyday use worldwide. North American interest has grown significantly, especially in recent years, and its use is now expressly mandated by a growing number of public-sector federal, state, provincial, and municipal agencies for their projects. The private sector is also invested in its use, with many developers and investors finding that its "functional" allocation of cost and description is far more easily understood by the layman than a traditional "trade/contract" allocation. This use is growing and is following the industry trend toward early involvement of constructors in the design process. With an increasing use of design–build and public–private partnership forms of procurement, contractors are actively exploring and adopting the concept of elemental cost management (planning, control, and analysis) to their advantage.

The ASTM E06.81 Building Economics subcommittee has been challenged by requests from industry practitioners to expand UNIFORMAT II into other forms of construction, specifically into the heavy and civil engineering sectors, and also into greater levels of detail. Meeting these challenges presents a number of issues, not least of which is the engendering of a fundamental understanding of what UNIFORMAT II is, how it is structured, how it is best used, and how it adds value to the construction industry.

Broad Principles in Elemental Classification

The goal is a system that can be applied at all stages of design development from planning through construction, using views that will be common in its hierarchical structure at all times, while based upon an ever changing/varying base of detail information.

THE CHALLENGE

- To represent an identifiable constructed functional entity or object
- To find a common view for a variety of design solutions in forms that can be compared one with another and applied in all instances of a particular entity
- To be of lasting value to industry users, be strictly structured, and rarely change
- To be modular and provide use in any project

THE FUNDAMENTAL CONCEPT

UNIFORMAT II elemental classification envisages elements as permanent physical components common within a constructed entity that perform particular functions regardless of their design or specification. The classification is a common thread that links activities and participants in any construction project from initial planning and implementation through operations, maintenance, and disposal.

When future construction work is estimated in the absence of any design, heuristic single-rate methods are often used (e.g., cost per unit of floor area applied to buildings). This is a broad-range tool that may get one into the right ballpark, but it lacks precision, and it might not be very project specific. Conversely, when all the design and construction documents are completed, a detailed and quite precise multirate construction estimate can be prepared. The first method enables the setting of a forecast figure prior to commencing design, and once this is completed the second method will determine whether that figure has been matched. The two methods may be appropriate in their respective phases of development, but neither is appropriate in the other phase. The discovery that completed design work does not match a client's budget is an expensive problem. To address this issue, cost-check estimates are introduced into the design process, and these will employ a wide variety of other estimating methods, each appropriate to the information available, in determining a more and more precise answer. This iterative process will account for the continually evolving design information, but it needs to permit meaningful comparison. The concept of functional elements is a valuable aid in this.

In summarizing cost back to a consistent, and fixed, set of elements, one does not need to change the estimating methods, just the way he or she views (sees) the result. Consistently viewing the evolving design by elements (functional elements) has proved to be a relatively simple, valuable, and successful solution.

Elemental cost analysis of completed work provides valuable data that can be fed forward into the modeling of future projects, replacing the single-rate method with a more nuanced multirate and project-specific budget projection.

UNIFORMAT II is about seeing construction as functional elements. Although developed for cost control during design, it has been found to be applicable for other uses too. These same elements may be used similarly to the chapters in a book, and so lend themselves to the structuring of reports and databases of all kinds in which the underlying detail is variable and differs from project to project.

Choosing appropriate elements that can be universally applied with the various forms of construction is not easy, and a good deal of rigor is required in their selection. Struck at too detailed a level, they will not be usable in the beginning, but at too high a level they will lack the granularity needed to achieve both planning and control. Finding a balanced middle ground is difficult.

In essence, the fundamental principle might also be described as one of "creating simplicity out of complexity" or "being able to see the wood for the trees" (i.e., having the ability to see the bigger picture without the noise of the details).

Functional Elements, Groups, and Tables

ELEMENTAL CLASSIFICATION

In UNIFORMAT II standard elemental classifications, a consistent structure is maintained. Through experience and common consent, the following arrangement has been found to work well.

Each elemental classification contains three hierarchal levels of aggregation:

- element,
- group element, and
- major group element.

These are further collectively grouped into tables that form the mandatory part of an ASTM standard classification.

One or more tables may be included within each elemental classification, and each elemental classification standard applies broadly to a common sector of the construction industry.

INDIVIDUAL ELEMENTS

The key to UNIFORMAT II, and other elemental classifications used worldwide, is the basic element (functional element). As applied here, the term *element* has a more nuanced and specific meaning than the regular dictionary definition.

As defined by ASTM International (ASTM E833 [5]):

> **element**, *n—in construction planning, design, specification, estimating, and cost analysis*, a significant component part of the whole that performs a specific function, or functions, regardless of design, specification, or construction method.
>
> DISCUSSION—While through analysis, or by direct application, construction estimates categorized into elements (functional elements) with allocated costs may be summarized in an elemental cost summary or elemental cost analysis; elements (functional elements) also provide a framework for consistent preliminary project description, outline, and performance specification, through all stages of planning, design, construction, and maintenance.

The selection of appropriate elements requires the melding of many factors before they may be selected as appropriate. Consideration of the following attributes, at a minimum, is necessary:

- Capable of being defined precisely
- Self-explanatory
- Of cost importance
- A functional component of the constructed entity
- Performs substantially the same function(s) irrespective of the constructed entity or its design solution
- Easily separable at all stages of development
- Quantifiable at all stages of development
- Capable of reconciliation with other elemental classifications
- To the extent possible, independent of other elements
- Allows comparison, estimate to estimate and project to project, in a meaningful way

In applying these attributes, an element will be unique, carry no burden of being tied to a specific form of construction, and be an instance of the stated function(s) that can take many physical forms. Consequently, it can, and will, be used repeatedly within the full UNIFORMAT II family of classifications.

An element's name can be seen as a short-form description of the function(s) that it performs and the data that will be collected, collated, stored, summarized, analyzed, referenced, and published in its name. An element might also be described as a container, within which relevant data, in numeric, textual, or visual form, or in a combination of these forms, can be recorded and maintained. The described content, magnitude, and construction method contained in an element will almost always differ from project to project and may vary from stage to stage within any project as it develops. It is that variance that needs to be controlled in order for a budget to be met.

Unfortunately, experience has shown that the philosophy of seeing function as an overall classification method is not well understood and consequently is frequently dismissed.

SUB-ELEMENTS

Although many seek further classification detail below the element, this is not included in ASTM elemental classifications other than by way of a broad description of content. Any sub-element classification is unlikely to ever be complete and will often be inappropriate for the level of design detail available at any specific point in time.

UNIFORMAT II will not create mandatory lower levels of detail, as this is counter to the intent and principle of viewing and summarizing the construction(s) with only as much detail as can be seen consistently in all stages of planning, design, construction, operation, maintenance, and disposal. This deliberate restriction also aids the view and transfer of information between users in different disciplines. It is a common base for all.

GROUPED ELEMENTS

Grouping is the gathering together of elements into relevant collections that further simplify the view of an entity.

As defined by ASTM International (ASTM E833 [5]):

> **group element**, *n—in construction planning, design, specification, estimating, and cost analysis*, a significant component part of the whole that includes relevant elements which, as a group perform a major specific function, or functions, regardless of design, specification, or construction method.

These group elements are then further gathered into the following, as defined by ASTM International (ASTM E833 [5]):

> **major group element**, *n—in construction planning, design, specification, estimating, and cost analysis*, a very significant component part of the whole that includes relevant group elements which, as a group perform a major

specific function, or functions, regardless of design, specification, or construction method.

Grouping elements caters to the human need for readability and simplicity by gathering together parts of a construction. These groupings may also be seen as milestone groupings in the process of construction. In any hierarchy, these groupings will generally be ordered in a typical construction sequence. They should also be backward-compatible to past (archaic) forms of construction and be suitable for existing structures as well as new.

Individual elements are typically not work packages, often used in work breakdown structures or contractually for delivering projects, although group and major group elements might well be.

These groupings are further gathered together into tables.

ELEMENTAL TABLES

To meet the intention that UNIFORMAT II be both structured and flexible in use, the grouped elements are formed into tables that represent a particular constructed entity—a bridge [6], a building [1], a road section, etc.

A fundamental principle in elemental cost analysis is that each constructed entity be individually identified and reported, to allow effective monitoring and control. To meet this requirement, elemental classification standards may contain one or more tables. Within each classification there may be both *core* tables (always required) and *optional* tables (applied in addition when needed) for each constructed entity.

Flexibility requires that each table be used as many times as there are individual functional entities that they represent within a project.

Structure requires that each table, when used, always be applied in its entirety, unchanged, with no ad hoc omissions or additions. This ensures that negative evidence is just as valuable/important/instructive as positive evidence. When viewed in a consistent summary form, any omission (intended or otherwise), and the relative magnitude of each element, is made more obvious and instructive. Gross omissions or a lack of design intent (by omission, error, or lack of understanding) can often be readily identified when this strict rule is maintained. "Cherry picking" from a list of elements precludes this simple check and is another reason that UNIFORMAT II classifications are published with discrete and separate tables.

As an example, ASTM E1557 [1] contains two tables, "Buildings" (Table I) and "Related Site Work" (Table II). Within any project, these two tables allow different forms of semi-independent construction, either large building/small site or vice versa, to be analyzed separately. Site size and site condition are not directly linked to building size, whereas building size will directly affect much of the content of the building table (Table I), and site size will affect most of the content of the site table (Table II). These are not necessarily directly or tightly linked or dependent as with roadway or bridge classifications, where sitework is more directly related to the size of the structure. Additionally, a building project with two buildings on one site will use **Table I** within ASTM E1557 twice (once for each separate building) and **Table II** within ASTM E1557 just once (for the site work surrounding both buildings).

Appendix A ("Bridge" [6]) and Appendix B ("Buildings and Related Sitework" [1]) included in this paper contain complete table structures to further aid in understanding this application principle.

PARENT–CHILD RELATIONSHIP

The parent–child relationship is adhered to throughout all forms of hierarchical classifications. This is especially true with UNIFORMAT II, as an individual element may be applied in various groups, which in turn are applied to major groups as well, within a variety of tables.

A simple example (using dot notation) is a foundation element that may be included in various groups in different tables and classifications.

- `Table/Classification.MajorGroupElement.GroupElement.`*`Element`*

Examples (extracted from Appendixes A and B):

- `Bridge.Substructure.Abutment.`*`Foundation`*
- `Bridge.Substructure.Pier.`*`Foundation`*
- `Building.Substructure.Foundation.`*`StandardFoundation`*

In each case the element remains functionally the same, but its content, description, magnitude, and construction method may differ depending upon what it is supporting.

The content of a `Bridge.Substructure.Pier.Foundation` element will differ markedly from that of a `Building.Substructure.Foundation` element, and in both these examples the content may differ again between long-span/short-span bridges or low-rise/high-rise buildings.

The differing data within each set of parent–child relationships will cause the analyzed data derived from each to differ. The element function will always remain the same; it is the solution that differs.

A generic representation of the UNIFORMAT II hierarchy is included here (**Fig. 1**) and demonstrates graphically this important parent–child hierarchical (nesting) relationship. It is a concept that is basic to all UNIFORMAT II classifications.

TABLES CONSIDERED AS MODULES

Every construction project is unique and differs in some way from others. Even ostensibly similar, or identical, projects will contain project-specific differences occasioned by external variables (e.g., location, timing, or ground conditions), in addition to any design or construction improvements that result from recent experience or new knowledge. As a basic principle, UNIFORMAT II tables do not individually attempt to encompass entire projects and are best considered as one of a set of building blocks that together form a particular project. As "modules," tables are used alone or in combination, as needed, to "build" any project. Being directed to consider, and control, each discrete elemental classification table as an individual

FIG. 1 Generic UNIFORMAT II parent-child relationship.

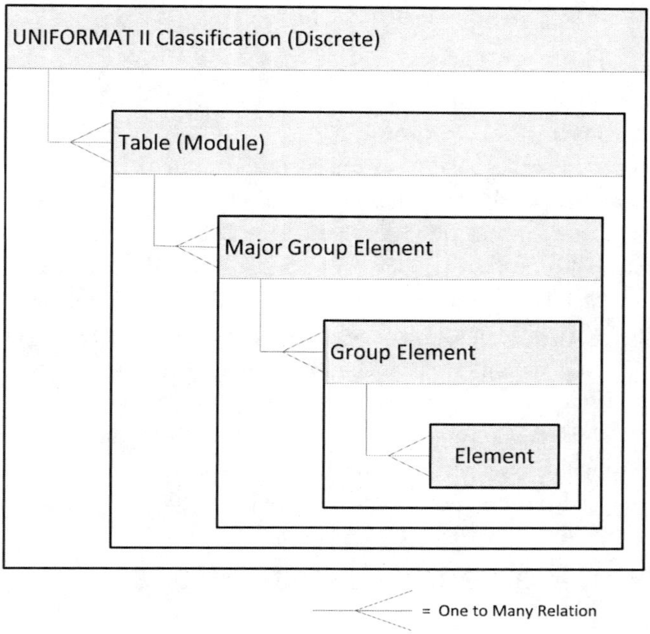

module focuses attention on the unique composition of each project and each entity. Employing UNIFORMAT II classification tables in a modular fashion creates the means for a directed, focused, and ultimately more successful management process. Doing so avoids unnecessary redundancy, and the concomitant temptation to omit elements in any system that considers the whole built environment as a single, large, lengthy classification. As noted above, each separate, discrete table/module may be used repeatedly, as often as needed. Consistent use within a project, and from project to project, makes for meaningful comparison; this creates added control and provides a valuable source of data that can be fed forward into future projects.

The hierarchical organization demonstrated thus far does continue beyond tables, but this is diverse and dependent upon the user's specific needs.

BEYOND TABLES

Tables can be further classified in a variety of ways appropriate to the users, their needs, and their intent. Adding a title or a designator to each repeated table will separate the tables and the details contained within them. Additional titles and designators might include the following:

- Location
- Date

- Type
- Phase
- Use
- Size
- Owner
- Designer
- Reference number

Any combination of these and other designators is also possible.

Any designators used for these further classifications are considered personal and specific to each client/practitioner's particular need and are too diverse to be reliably documented in a fixed classification standard. Avoiding such prescriptive detail within any standard classification maintains a further degree of classification flexibility in use, and yet retains a very specific level of consistency and rigor within the table classification and on down though the range of groups and elements that a table includes.

The parent–child relationship described above is an important principle and applies throughout the classification process. As stated, an element may be used several times in a classification table, but each time with a different parent.

A Family of Elemental Classifications

TABLES IN USE

Each UNIFORMAT II classification standard represents a broad industry sector providing tables of elements that are tailored and best suited for use in that sector. Each classification is intended to be discrete and does not include work that is typically outside its industry sector.

Just as each individual element is based on an end function, the existing and proposed UNIFORMAT II elemental classification standards, although broadly based on construction industry sectors, are based on function too. The current ASTM E06.81 Subcommittee UNIFORMAT II Task Group's overall guidelines [7] include the graphic in **Fig. 2**, which is presently serving as a master plan.

This master plan, while not set in stone, has provided the useful function of directing attention to the development of specific, discrete classifications that include the following examples:

- Dam and Canal (DAMC)—includes dams, canals, locks, levees, and reservoirs (essentially those constructed entities that contain, channel, or control surface water). Primary function: control/contain surface water.
- Transportation Surface (TRAN)—includes roadways, railways, runways, taxiways, and aprons (essentially all those constructed entities that aid, support, channel, or direct mechanical transportation devices at grade level). Primary function: carry/support moving transport loads.
- Bridge (BRDG)—includes bridge constructions that carry transportation surfaces and other entities (pipes and pipelines) over and across natural or man-made obstacles. Primary function: provide access across/over obstacles.

FIG. 2 Classification master plan.

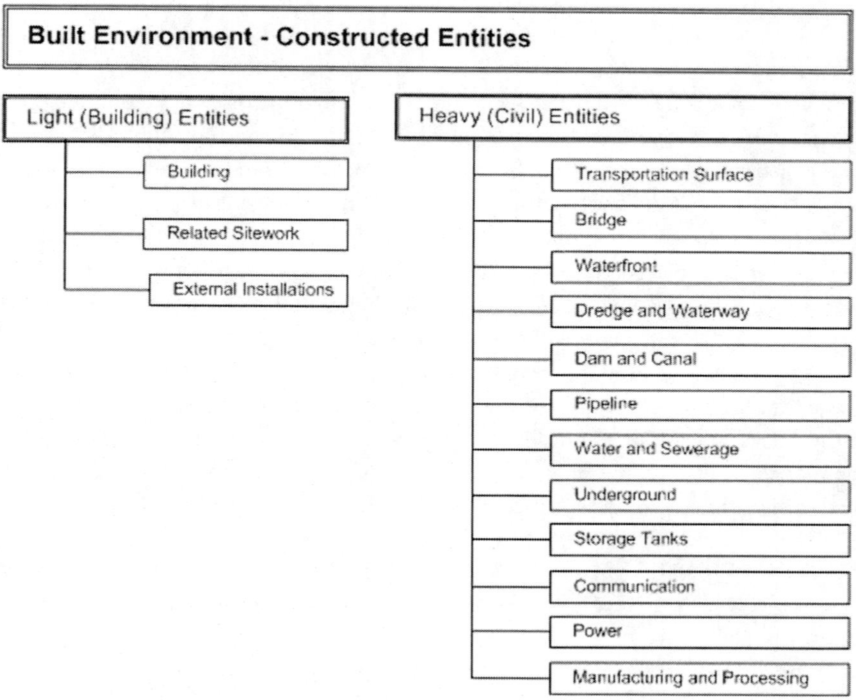

- Underground (UGRD)—includes tunnels, shafts, subways, and mines that require or provide access. Excludes entities (utilities) that are buried. Primary function: provide access through/beneath obstacles.
- Waterfront (WTFR)—includes wharfs, piers, jetties, docks, breakwaters, and groynes (groins) (essentially all those constructed entities at the interface of land and water created for the purpose of facilitating an exchange between waterborne and land-based transportation, maintaining controlled access to aquatic areas, or providing erosion control and protection). Primary function: protect/facilitate land/water access.
- Water and sewerage (WTRS)—includes potable water supply (wells, pipelines, and treatment plants) and sewerage (waste water trunk sewers, treatment plants, treatment ponds, and anti-pollution facilities). Excludes reservoirs and storage tanks. Primary function: supply and treat liquids.
- Building (BLDG)—includes residential, educational, cultural, health, religious, government, public, commercial, transportation (road, rail, air, and water), industrial, recreation and entertainment, scientific research, and farm and animal buildings. Excludes related sitework and other external installations and entities. Primary function: provide shelter.

- Related sitework (RSWK)—includes light construction work that is related to the building(s) and within specific (or immediate) site boundaries (see ASTM E1557 for a more complete description of inclusions and exclusions). Excludes external installations. Primary function: provide support.

Buildings are a distinct industry sector, as are other entities such as bridges, roads, and tunnels. These distinct sectors may apply industry-specific terminology or use a particular range of materials, in different quantities, and in a variety of ways (construction methods), in a blend that can be unique to their sector. Consequently the range of outcomes (data) within a common element may be wide, and for this reason the separation of elemental data by sector and function is important.

We should also recognize the fine distinction being drawn between *function* and *use* that is being applied here. Included in the examples above is a stated primary function that the entity will typically accomplish. A building's primary function is to provide shelter, but it may be used for many different purposes that may change through its lifetime. This is not an uncommon occurrence, as we repurpose existing structures. Although the "Building" table will remain consistent at all times, practitioners may wish to apply added classification distinctions between uses, as noted earlier when we discussed parent–child relationships ("Beyond Tables").

ASTM'S UNIFORMAT II FAMILY OF STANDARDS

The published UNIFORMAT II classifications of discrete constructed entities are currently the following:

- ASTM E1557 on building elements and related sitework [1]
- ASTM E2103 on bridge elements [6]

In preparation are classifications for vehicular surfaces (roadways, railways, and runways), underground structures (tunnels), waterfront structures (jetties, piers, docks, and defenses), and water and waste water treatment elements. Because these must adequately meet the criteria for philosophy, logic, and use, as outlined above, their drafting is necessarily a slow and painstaking process.

Elemental classifications of constructed entities are not the sole content of the UNIFORMAT II family of standards, either. There are additional published standard classifications that contain tables (modules) for the additional costs essential in the preparation of elemental cost estimates and budgets.

- ASTM E2083 on building construction field requirements and office overhead and profit [8]
- ASTM E2168 on allowance, contingency, and reserve sums [9]
- ASTM E2516 on cost estimate classification systems [10]
- ASTM E2620 Program and Project Estimate Summaries [11]

Nor is the family confined to classifications alone. There also are standard practices that relate to the cost-management process and presentation format.

- ASTM E1804 on performing and reporting cost analyses during the design phase of a project [12]
- ASTM E2514 on presentation formats of elemental cost estimates, summaries, and analyses [13]

Family Uses

Although it evolved to deal with cost management (cost planning, cost control, and cost analysis), the same hierarchical structure has been found useful for other activities also.

THE USERS OF UNIFORMAT II INCLUDE

Financial and investment—typically by owners, developers, bankers, lenders, accountants, and financial managers.

Implementation—primarily project managers; facilities programmers; designers, including architects and engineers; and project control specialists, including cost planners, estimators, schedulers, specification writers, and risk analysts.

Facilities management—property portfolio managers, operating staff, and maintenance staff.

Others—additional users may be manufacturers, educators, students, and other project stakeholders.

UNIFORMAT II classification should be applied when undertaking the following work on constructed entities:

FINANCING AND INVESTING

Structuring costs on an elemental basis for economic evaluations (ASTM E917 [14], ASTM E964 [15], ASTM E1057 [16], ASTM E1074 [17], ASTM E1121 [18], and ASTM E1804 [12]) early in the design process helps reduce the cost of early financial analysis and can contribute to substantial design and operational savings before decisions have been made that limit options for potential savings.

IMPLEMENTATION

Cost modeling; cost planning; and estimating and controlling project time and cost during planning, design, and construction—use UNIFORMAT II classifications to prepare budgets and to establish elemental cost plans before design begins. Project managers and project control specialists use these cost plans as a standard against which to measure and control project cost and quality and to set design-to-cost targets.

Conducting value engineering workshops—use UNIFORMAT II as a checklist to ensure that alternatives for all elements of significant cost in the building project are analyzed in the creativity phase of the job plan (ASTM E1699 [19]). Also, use the elemental cost data to expedite the development of cost models for building systems.

Develop initial project master schedules—because projects are essentially built element by element, UNIFORMAT II classifications are an appropriate basis for preparing construction schedules at the start of the design process. Project managers and project control specialists use these time plans as a baseline against which to measure and control project time and to set milestone target dates.

Performing risk analyses—simulation is one technique (ASTM E1369 [20]) for developing probability distributions of building costs when evaluating the economic risk in undertaking a building project. Use individual elements and group elements in UNIFORMAT II for developing probability distributions of elemental costs. From these distributions, build up probability distributions of total project costs to establish acceptable project contingencies or to serve as inputs to an economic analysis. (See ASTM E1185 [21] for guidance on which economic method to use.)

Structuring preliminary project descriptions during the conceptual design phase—this facilitates the description of the scope of the project for the client in a clear, concise, and logical sequence; it provides the basis for the preparation of more detailed elemental estimates during the early concept and preliminary design phases, and it enhances communications among designers and other building professionals by providing a clear statement of the designer's intent. A useful guide to this process is published by the Construction Specifications Institute [22].

Coding and referencing standard details in computer-aided design systems—this allows an architect, for example, to reference an exterior wall assembly according to UNIFORMAT II element designations and build up a database of standard details structured according to the classification. This is particularly appropriate for design modeling and building information modeling applications.

MANAGING FACILITIES

Recording and writing property condition assessment reports in a structured way, using UNIFORMAT II classifications, provides for a consistent, accessible, and searchable database of real property inventory.

OTHER ACTIVITIES

One can structure cost manuals and record construction, operating, and maintenance costs in a database. Having a manual or database in an elemental format greatly assists the performance economic analysis early in the design stage, and at reasonable cost.

Caveat Emptor

Through several years of reading, writing, and discussing the elemental technique, some common misconceptions have become readily apparent. These are no doubt the result of a multitude of reasons. Applying the elemental technique to an environment in which we are each wedded to our profession and to ideas we learned in training and continue to apply every day does require some mental adjustment.

WHAT UNIFORMAT II IS NOT

- A new way of estimating or contracting. It does not change the traditional way the construction industry presently operates; it simply provides an alternative way of summarizing and analyzing the work in a manner that will be usable through all the life-cycle stages of construction. Only by comparing like with like can one make meaningful comparisons.
- An ad hoc collection of cost or descriptive headings.
- A work breakdown structure (WBS), as has been stated by many; if it is a WBS, then it is a nuanced one, toward cost management (cost planning, cost

control, cost analysis). The Project Management Institute has published a very specific definition of a WBS [23] that applies to a hierarchy of items that is both variable in content and inclusive of work that is in addition to construction. UNIFORMAT II is better described as an elemental cost breakdown structure, or perhaps more broadly as an elemental breakdown structure.

WHAT UNIFORMAT II IS

- Flexible in use.
- Structured.
- Usable by many disciplines.
- A summary tool.
- A control tool.
- Usable for both mathematical and text-driven purposes, retaining a common classification for both.
- Applied to construction components.
- Applied as a form of parametric estimation, in which analyzed elemental data from past projects are fed forward into new projects at the early planning stage, lending itself well to cost modeling in the early planning phase when a variety of options can be modeled in a quite project-specific way.
- An excellent cost and descriptive structure on which to base value engineering workshops.

Conclusion

UNIFORMAT II is, overall, a structured framework within which elemental classification tables can be used individually or as a module in conjunction with other elemental classification tables.

Elemental classification is a simple, widely usable, alternate way of viewing, reporting, and controlling construction entities, from initial idea to end of life. The concept is applied worldwide and has been in use for many years with significant success and widespread adoption.

The UNIFORMAT II framework is planned to grow—and is growing—into a family of related elemental classifications. Each individual classification is intended to classify the component elements of a broad construction sector in a manner that is

- logical,
- consistent,
- structured, and
- usable.

Doing so permits

- repeatability in use,
- reduced redundancy in content,
- use by a range of professional disciplines (throughout the whole planning, design, construction, operation, and maintenance life cycle), and
- valuable analysis, data reporting, comparison, control, recording, and data retention.

Each classification is

- discrete and focused on one construction's function, and yet
- sufficiently generic to include the many end uses of each construction.

This is accomplished through the following:

- specific rules,
- precise definition,
- common presentation, and
- knowledgeable application.

Appendix A

An example ONE TABLE classification from ASTM E2103-11 [6].

TABLE 1. Bridge Elements - UNIFORMAT II

Level 1 Major Group Elements	Level 2 Group Elements	Level 3 Individual Elements
Substructure	Piers	Foundations
		Walls
		Columns
		Cap Beams
	Towers	Foundations
		Walls
		Columns
		Cap Beams
	Abutments	Foundations
		Stem Abutments
		Wing Walls
	Other Supports	Thrust Blocks
		Anchorages
Superstructure	Short Span Assemblies	Flexural Members
		Diaphragms
		Bracings
		Bearings
	Long Span Assemblies	Ribs
		Cables
		Hangers and Spandrels
		Ties
		Truss Members
		Segmental Box Girders
	Deck	Structural Surface
		Wearing Surface

TABLE 1. *(Continued)*

Level 1 Major Group Elements	Level 2 Group Elements	Level 3 Individual Elements
Protection	Structure Protection	Slope Walls
		Expansion Joints
		Protective Coats
		Sacrificial Beams
		Drainage Systems
		Inspection and Maintenance Systems
	Traffic Protection	Barriers
		Protective Shields
		Traffic Controls
	Other Protection	Lighting
		Signage
		Sound Barrier Walls
		Air Pressure Barriers
		Enclosure
Site work	Site Preparation	Clearing and Grubbing
		Demolition and Relocation
		Earthwork
		Hazardous Material Handling
		Environmental Restoration/Replacement
	Approach construction	Approach Slabs
		Sleeper Slabs
		Earth Retention Systems

Appendix B

An example TWO-TABLE classification from ASTM E1557-09 [1].

TABLE 2 Building Elements - UNIFORMAT II

Level 1 Major Group Elements	Level 2 Group Elements	Level 3 Elements
Substructure	Foundations	Standard Foundations
		Special Foundations
		Slab on Grade

TABLE 2 *(Continued)*

Level 1 Major Group Elements	Level 2 Group Elements	Level 3 Elements
	Basement Construction	Basement Excavation
		Basement Walls
Shell	Superstructure	Floor Construction
		Roof Construction
	Exterior Enclosure	Exterior Walls
		Exterior Windows
		Exterior Doors
	Roofing	Roof Coverings
		Roof Openings
Interiors	Interior Construction	Partitions
		Interior Doors
		Fittings
	Stairs	Stair Construction
		Stair Finishes
	Interior Finishes	Wall Finishes
		Floor Finishes
		Ceiling Finishes
Services	Conveying	Elevators & Lifts
		Escalators & Moving Walks
		Other Conveying Systems
	Plumbing	Plumbing Systems
		Domestic Water Distribution
		Sanitary Waste
		Rain water Drainage
		Other Plumbing Systems
	HVAC	Energy Supply
		Heat Generating Systems
		Cooling Generating Systems
		Distribution Systems
		Terminal & Package Units
		Controls & Instrumentation
		Systems Testing & Balancing
		Other HVAC Systems & Equipment
	Fire Protection	Sprinklers
		Standpipes
		Fire Protection specialties
		Other Fire Protection Systems

TABLE 2 *(Continued)*

Level 1 Major Group Elements	Level 2 Group Elements	Level 3 Elements
	Electrical	Electrical Service & Distribution
		Lighting and Branch Wiring
		Communications & Security
		Other Electrical Systems
Equipment and Furnishings	Equipment	Commercial Equipment
		Institutional Equipment
		Vehicular Equipment
		Other Equipment
	Furnishings	Fixed Furnishings
		Movable Furnishings
Special Construction & Demolition	Special Construction	Special Structures
		Integrated construction
		Special Construction Systems
		Special Facilities
		Special Controls and Instrumentation
	Selective Building Demolition	Building Element demolition
		Hazardous Component Abatement

TABLE 3 Building Sitework Elements - UNIFORMAT II

Level 1 Major Group Elements	Level 2 Group Elements	Level 3 Elements
Building Sitework	Site Preparation	Site Clearing
		Site Demolition and Relocation
		Site Earthwork
		Hazardous Waste Remediation
	Site Improvements	Roadways
		Parking Lots
		Pedestrian Paving
		Site Development
		Landscaping
	Site Mechanical Utilities	Water Supply
		Sanitary Sewer
		Storm Sewer
		Heating Distribution
		Cooling Distribution

TABLE 3 *(Continued)*

Level 1 Major Group Elements	Level 2 Group Elements	Level 3 Elements
		Fuel Distribution
		Other Site Mechanical Utilities
	Site Electrical Utilities	Electrical Distribution
		Site Lighting
		Site Communications and Security
		Other Site Electrical Utilities
	Other Site Construction	Service and Pedestrian Tunnels
		Other Site Systems & Equipment

References

[1] ASTM E1557-09: Classification of Building Elements and Related Sitework—UNIFORMAT II, *Annual Book of ASTM Standards*, ASTM International, West Conshohocken, PA, 2012.

[2] Hanscomb Associates Inc., *Automated Cost Control and Estimating System*, General Services Administration, Washington, D.C., 1975.

[3] BCIS, *Standard Form of Cost Analysis*, Royal Institution of Chartered Surveyors, London, UK, 1969.

[4] Royal Institution of Chartered Surveyors (RICS), *RICS New Rules of Measurement, Order of Cost Estimating and Elemental Cost Planning*, RICS, Coventry, UK, 2009.

[5] ASTM E833-13: Terminology of Building Economics, *Annual Book of ASTM Standards*, ASTM International, West Conshohocken, PA, 2013.

[6] ASTM E2103-11: Classification for Bridge Elements—UNIFORMAT II, *Annual Book of ASTM Standards*, ASTM International, West Conshohocken, PA, 2012.

[7] ASTM International, 2012, "E06 Committee Documents," http://myastm.astm.org (Last accessed 26 March 2014).

[8] ASTM E2083-05: Classification for Building Construction Field Requirements, and Office Overhead & Profit, *Annual Book of ASTM Standards*, ASTM International, West Conshohocken, PA, 2012.

[9] ASTM E2168-10: Classification for Allowance, Contingency, and Reserve Sums in Building Construction Estimating, *Annual Book of ASTM Standards*, ASTM International, West Conshohocken, PA, 2013.

[10] ASTM E2516-11: Classification for Cost Estimate Classification System, *Annual Book of ASTM Standards*, ASTM International, West Conshohocken, PA, 2013.

[11] ASTM E2620-08: Classification for Program and Project Estimate Summaries, *Annual Book of ASTM Standards*, ASTM International, West Conshohocken, PA, 2013.

[12] ASTM E1804-12: Practice for Performing and Reporting Cost Analysis during the Design Phase of a Project, *Annual Book of ASTM Standards*, ASTM International, West Conshohocken, PA, 2012.

[13] ASTM E2514-08: Practice for Presentation Format of Elemental Cost Estimates, Summaries, and Analyses, *Annual Book of ASTM Standards*, ASTM International, West Conshohocken, PA, 2013.

[14] ASTM E917-05: Practice for Measuring Life-cycle Costs of Buildings and Building Systems, *Annual Book of ASTM Standards*, ASTM International, West Conshohocken, PA, 2012.

[15] ASTM E964-06: Practice for Measuring Benefit-to-cost and Savings-to-investment Ratios for Buildings and Building Systems, *Annual Book of ASTM Standards*, ASTM International, West Conshohocken, PA, 2012.

[16] ASTM E1057-06: Practice for Measuring Internal Rate of Return and Adjusted Internal Rate of Return for Investments in Buildings and Building Systems, *Annual Book of ASTM Standards*, ASTM International, West Conshohocken, PA, 2012.

[17] ASTM E1074-09: Practice for Measuring Net Benefits and Net Savings for Investment in Buildings and Building Systems, *Annual Book of ASTM Standards*, ASTM International, West Conshohocken, PA, 2012.

[18] ASTM E1121-12: Practice for Measuring Payback for Investments in Buildings and Building Systems, *Annual Book of ASTM Standards*, ASTM International, West Conshohocken, PA, 2012.

[19] ASTM E1699-10: Practice for Performing Value Analysis (VA) of Buildings and Building Systems and Other Constructed Projects, *Annual Book of ASTM Standards*, ASTM International, West Conshohocken, PA, 2012.

[20] ASTM E1369-11: Guide for Selecting Techniques for Treating Uncertainty and Risk in the Economic Evaluation of Buildings and Building Systems, *Annual Book of ASTM Standards*, ASTM International, West Conshohocken, PA, 2012.

[21] ASTM E1185-12: Guide for Selecting Economic Methods for Evaluating Investments in Buildings and Building Systems, *Annual Book of ASTM Standards*, ASTM International, West Conshohocken, PA, 2012.

[22] Construction Specifications Institute, *PPDFormat: A Guide for Developing Preliminary Project Descriptions*, The Construction Specifications Institute, Alexandria, VA, 2010.

[23] Norman, E. S., Brotherton, S. A., and Fried, R. T., *Work Breakdown Structures*, John Wiley & Sons, Hoboken, NJ, 2008, pp. 5–7.

STP 1586, 2014 / available online at www.astm.org / doi: 10.1520/STP158620140033

Perry Daneshgari[1] and Heather Moore[2]

Feedback from the Source Improving Productivity on Construction Jobsites

Reference

Daneshgari, Perry and Moore, Heather, "Feedback from the Source Improving Productivity on Construction Jobsites," *Ideas to Impact: How Building Economic Standards Keep You on Track*, STP 1586, Robert E. Chapman and Michael N. Goodkind, Eds., pp. 102–126, doi:10.1520/STP158620140033, ASTM International, West Conshohocken, PA 2014.[3]

ABSTRACT

The construction industry relies on skilled trade craftsmen to transfer a concept from design to a physical built entity. In addition, construction jobsites are temporary workspaces. These two characteristics classify construction as a complex production, in which knowledge and experience are gained locally by individuals in the work environment. It also means that workers must adapt and coordinate in the local work environment in order to produce. Without a correct measurement system, the adaption and coordination and resulting information gained are invisible outside of the jobsite, which costs construction contractors and contributes to over $\$10 \times 10^9$ of unnecessary cost to the construction industry annually. ASTM's Standard Practice for Job Productivity Measurement (ASTM E2691-11) was developed to provide a measurement of task, project, and industry productivity. The practice provides ongoing and instantaneous feedback from the source of work on construction jobsites, which are the craftsmen. MCA, Inc. originally developed and implemented this measurement system over fifteen years ago and has seen results of minimum 20 % improvement in productivity on jobsites across the United States and Canada. ASTM E2691 describes how productivity is measured, monitored, and analyzed, and leads to other by-products of information gathered from the source. In addition, ASTM has

Manuscript received March 31, 2014; accepted for publication June 13, 2014; published online August 14, 2014.

[1]President and CEO, MCA, Inc., 363 East Grand Blanc Road, Grand Blanc, MI 48439, e-mail: perry@mca.net

[2]Vice President of Operations, 363 East Grand Blanc Road, Grand Blanc, MI 48439, e-mail: hmoore@mca.net

[3]ASTM Symposium on *Ideas to Impact: How Building Economic Standards Keep You on Track* on October 5, 2014 in New Orleans, LA.

published Manual 65 on how to use ASTM E2691 with several case applications. This paper explains how ASTM E2691 can be used by all stakeholders of the construction industry supply chain. The standard is explained and examples given for how it can be used by each stakeholder. In addition, a model of information entropy is introduced based on data gathered using JPM, leading to a new approach that can be used to reduce information entropy and improve the time, cost, and quality of construction.

Keywords

construction productivity, productivity, productivity differential, ASTM E2691, job productivity measurement, value transfer, labor productivity, information entropy, information modeling, statistical process control

Introduction

The construction industry is the backbone of a vital resource for human life. Beyond providing basic shelter, the built environment is necessary for commerce as we know it today. If the translation of inputs to outcome in construction can be improved through improved productivity, it will have a significant impact on human life. ASTM Subcommittee E06.81 for Building Economics has contributed the Standard Practice for Job Productivity Measurement (ASTM E2691) [1] to this effort, by providing a means for measuring productivity at the task, project, and industry level. The subcommittee has also produced Manual 65 on how to use E2691 [2]. Other industries such as manufacturing that have undergone "industrialization" have followed this path by finding a method of measuring work and productivity that relies on the source of information about the work: skilled trade. Construction has surpassed the proverbial "fork in the road" of finding a new way of producing that requires fewer resources. There are many examples of companies, projects, and individuals that are finding ways of doing work by translating the knowledge of the work into better ways of physically producing. The principles of job productivity measurement (JPM) were designed to allow the international construction industry to improve its productivity for all users by relying on proven principles learned from construction and other industries.

Trends in the Construction Industry

The construction industry is a major contributor to the U.S. economy. Its contribution to the U.S. gross domestic product (GDP) can be as high as 10 % or as low as 5 % [3]. Construction worldwide contributes up to 6.5 % to the World Gross Product [4]. The U.S. construction market is 15 %–25 % of the entire world's construction market. Every percent of improvement in productivity in the U.S. construction industry equates to more than $\$10 \times 10^9$ dollars and will have a major effect on the national economy and resource usage footprint.

Productivity in construction has been shown to lag behind other industries over the last three decades [5] (**Fig. 1**). Construction productivity cannot be

FIG. 1 Productivity trend in construction compared to other industries [5].

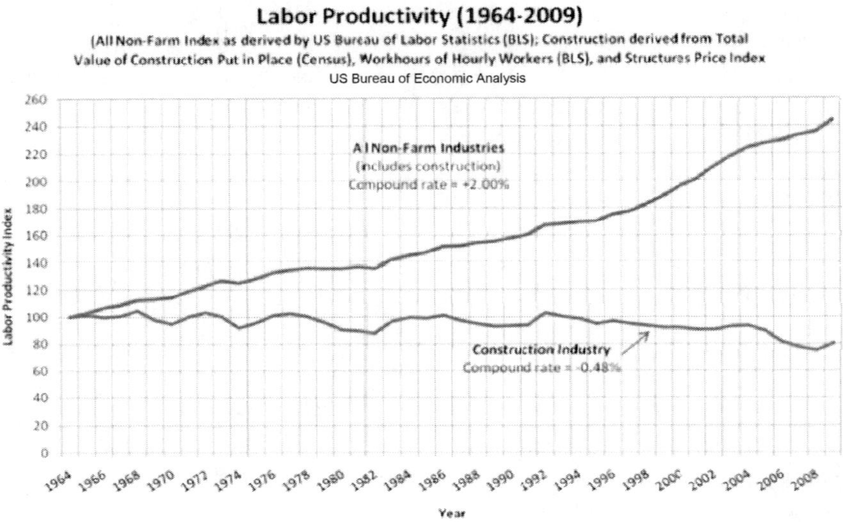

improved to be at par with other industries unless it is reliably measured, and the measurements taken acted upon for improvement.

The lack of reliable measurement methods and standards led the National Institute of Standards and Technology (NIST) to turn to ASTM to develop a standard for job productivity measurement in construction. ASTM E2691 [1] was developed to measure construction productivity at three levels: task, project, and industry. The ASTM standard practice periodically and continuously informs key stakeholders of productivity changes, which allows them to take corrective action. By measuring productivity and its changes during a construction project's progress, the issues can be resolved in a timely manner, leading to reductions in waste and rework.

The National Research Council launched a committee in 2008 to study how to improve and advance competitiveness and productivity in the construction industry. The committee found that one obstacle to improving productivity and competitiveness was the variation in standards, requirements, and processes required by different types of construction and construction projects, as well as the lack of an industry-wide strategy to improve productivity. In addition, the lack of effective performance measurements for construction-related tasks, projects, and the industry were recognized as obstacles to improvement [6]. The JPM standard [1] helps address that need.

While the productivity in the manufacturing industry has improved by 400 % over the last century, the construction industry's productivity has, in the best case, stayed flat or turned negative (**Fig. 1**). One main reason for the improvement of the manufacturing and other industries' productivity is the "industrialization" of those industries. The improvements seen in agriculture, manufacturing, and trade were made possible because the industries were able to produce more physical output with the same knowledge input of the skilled trade. In other words, the reliance on highly trained and highly skilled workers for their hands shifted to relying on them for their knowledge of the work. This started with Fredrick Taylor in 1911 and continued with Frank and Lilian Gilbreth, Henry Ford, Dr. Shewhart, Dr. Deming, and Taiichi Ohno [7]. Industrialization of any industry will rely on the following five factors:

1. Management of labor
2. Management of work
3. Lean operations
4. Modeling and simulation
5. Feedback from the source

The construction industry is just embarking on steps one and two by understanding how to manage the labor beyond just reliance on skilled trade. Lean operations and a "one best way" will be the future of the industry once the tacit knowledge of the trades can be made explicit and used to improve efficiency and effectiveness on jobsites. One of the basic assumptions of the development of JPM was to assure ability of the feedback from the source. This paper explores that hypothesis. In addition, recent research has been used to demonstrate and verify the criticality of this input for improving productivity in construction.

Productivity Measurement in Construction and Other Industries

Despite the challenges to measuring productivity, several attempts have been made over the last 60 years to measure and improve productivity. The methods used can be categorized into two general methods for measuring productivity: economic-based and industrial engineering (IE)-based. Economic measurements look at the inputs and outputs to systems while IE measurements look at the processes within the system translating the inputs to outputs. JPM combines these two approaches and, in addition, relies on feedback from the source of information about the work—the worker—to report on construction outcome. Feedback from the source has been a key contributor to industrialization in other industries. When skilled trade is the source of information about the work, the measurement of work must use the skilled trades' input from the way they see the work. Any other attempts to quantify work such as hours, dollars, or quantities will fall short of a true reflection of the work. Without one best way to-date in construction, the work has to be measured against a reference point of the way the work is conducted.

A review of typical economic and IE-based measurements shows that there have been many attempts to quantify and measure productivity in construction.

From the standpoint of economic measurements, various sources of input and output data include R.S. Means, Construction Industry Institute (CII), and the Bureau of Labor and Statistics (BLS) [8]. Oglesby et al. [9] published that "productivity....is measured primarily in terms of cost....good productivity is work accomplished at a fair price to the owner and at a reasonable cost to the contractor." They suggested work sampling and questions and interviews to measure productivity, but still define it economically in terms of cost. Park et al. [10] took a different approach to economic input/output measurement, using industry input to define direct and indirect accounts of construction activities. Indirect activities such as cleanup, demolition, and testing were not included as inputs to the economic productivity calculation. Although this method attempts to capture labor activity in the measurement, it is not real-time, and relies on input from construction professionals rather than field labor.

The second general method for measuring productivity is IE-based measurements. IE measurements such as time and motion studies focus on observing, measuring, and improving workers' environments. Early IE pioneers claimed that where a time and motion study was applied, worker output doubled [11]. IE-based methods in the literature include work sampling, and labor impact studies. Adrian [12] first identified the need for direct observation and sampling from construction jobsites to measure productivity. Thomas and Yiakoumis [13] introduced a "factor model" for measuring construction productivity, which was a first attempt to compare the completion of construction to the effort expended, whereas previous methods only used quantities installed or hours worked to denote completion. Allmon et al. [8] studied 72 projects in Austin, TX over a period of 25 years using work sampling, categorizing activities into direct work, supportive work, and delays. Dai et al. [14] studied factors impacting labor productivity by quantifying input from workers. They triangulated the results by correlating the impact factors with crews' perceptions of their project's productivity, and found that the presence of negative impact factors had a negative impact on job productivity. Noor [15] conducted studies using the daily visit method, which requires daily site visits to observe work completed through engineering drawings to show completion and any notations of causes for delay beyond 15 min. Thomas et al. [16] also attempted to measure "disruptions" observed on masonry projects in seven different countries. However, the intention of this method was to remove the disruptions from the data on productivity measurement so project-to-project benchmarks could be established. In this case, the disruption data was not used to understand and resolve the causes for disruption.

Recently, simulation methods have been used to understand job productivity impacts. For example, activity-based modeling (ABM) has been used to understand the real-time impacts of various jobsite scenarios [17–19]. However, the model parameters do not rely on field labor input, and have not been verified with real job results [20].

The literature suggests that studies of real-time field labor situations at construction jobsites, which would provide feedback from the source of information about how the work is happening, have not been taken into account in productivity measurement. Although Dai et al. [14] emphasize the importance of the craft worker input, and quantify labor's perception of productivity impacts in terms of severity and frequency, the data is still lacking direct observation of the work. However, their results support the validity of field labor input in productivity measurement. Choy and Ruwanpura [21] come closest to studying the crew-level obstacles, arguing that productivity in construction operations is due to so-called "situations," which include unexpected events, and other factors. They did not find any historical real-time or ongoing measurements of these impacts on projects. This level of study is needed to understand what information is lost from the point of installation.

In summary, **Fig. 2** shows the culmination of where productivity studies have focused to-date in construction [22,23]. The top portion of the figure shows a simplistic model of the inputs to construction jobsites, with the output being construction put in place. The numbers label each element, and the table in the bottom left hand corner of the figure indicates which elements have been measured with which studies. The information from the source as defined above exists within the

FIG. 2 Focus of state-of-the-art productivity studies [23].

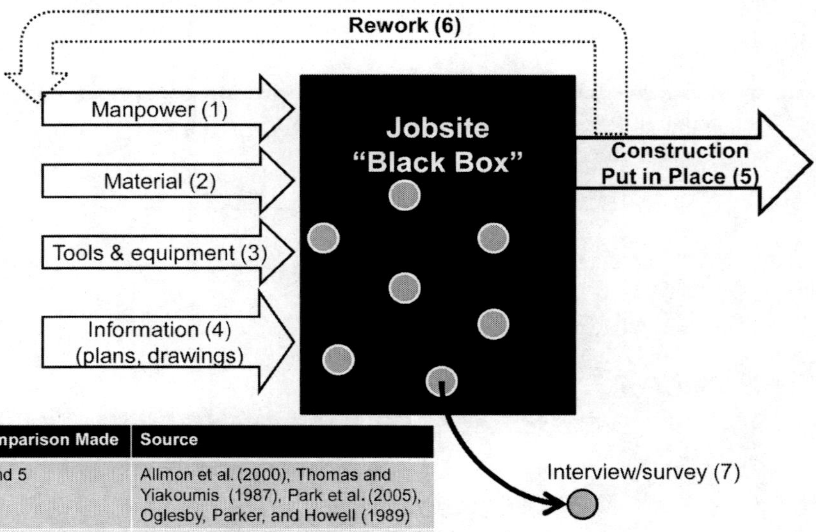

Comparison Made	Source
1 and 5	Allmon et al. (2000), Thomas and Yiakoumis (1987), Park et al. (2005), Oglesby, Parker, and Howell (1989)
1 and (5 minus 6)	Thomas et al. (1992)
5 and 7	Dai et al. (2009), Noor (1998)
4 and 7	Adrian (1995), Ballard et al. (2002)

"jobsite black box," but can only be captured within the box itself. In other words, the information propagation occurring on the jobsite is what this research is intended to study. **Figure 2** shows that the state-of-the-art studies of productivity tend to focus on the inputs to and outputs from the jobsite only. Studies that include direct input field labor (showing in **Fig. 2** as the "gray matter" within the black box) only consist of interviews or surveys away from the work environment, rather than real-time study of how the field labor interactions are occurring and what information is propagated.

Work is effort expended over duration of task. It has many dimensions and is expended non-linearly. Typical measurements of productivity measure effort expended for a given duration or relative to the number of units installed. Using these results to measure the project assumes the relative impacts or weights of activities, or "cost codes" as are sometimes used, remain the same throughout the life of a project. This represents the work as shown in the model presented in **Fig. 3**. However, this linear model does not match what happens in the construction environment. The work ramps up and down during the life of the project, as shown in **Fig. 4**.

Equation 1 shows what the measurement of work would be associated with **Fig. 3**, and Eq 2 is a more accurate reflection of the ongoing and instantaneous measurement of work, as shown in **Fig. 4**.

$$W = e \times d \tag{1}$$

$$W = \int_{t=0}^{t=n} (e_t \times \partial d_t) \tag{2}$$

FIG. 3 Work represented as sum of work associated with cost codes, assuming linear effort per cost code for the project duration.

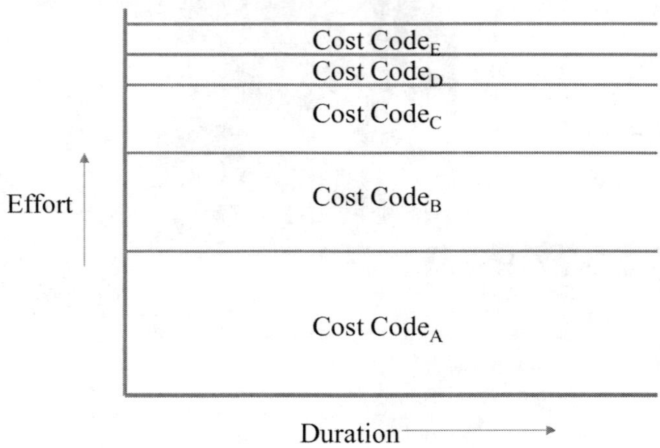

FIG. 4 Work represented as a non-linear curve of increase and decrease in effort over the project duration.

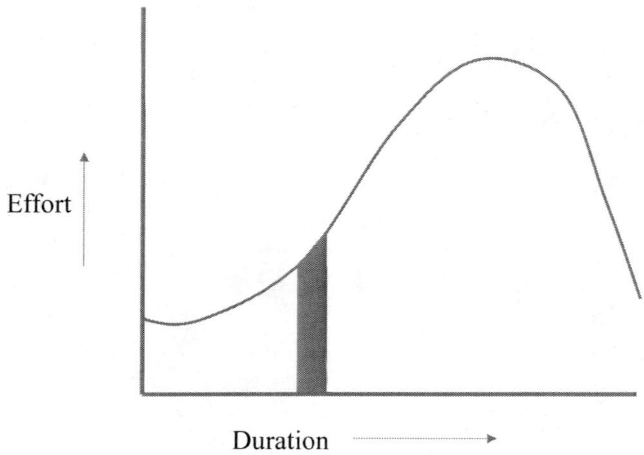

where:

e = effort (man-hours),

d = duration of project,

t = individual unit of time on the project, and

n = final project duration.

In addition to the non-linearity of the work on a project, activities come and go throughout the life of the project, making different contributions at different times, the amount of which is truly dependent on the effort, duration and difficulty of activities. The work associated with those activities is dynamic and non-linear. A model as shown in **Fig. 5** accounts for the fact that the work itself plus its components, shown in this example as cost codes, change constantly as a job progress. For instance, the effort to install one segment of 50 ft (15.2 m) of pipe is not necessarily equivalent to the effort to install the next segment of 50 ft (15.2 m) of pipe. If the first 50 ft (15.2 m) of pipe has more bends, welds, cuts, or other work involved with it, or is in a more congested area or smaller space, the effort involved for that segment of the work will be higher. This difficulty of installation and work breakdown is known by the skilled trade and needs to be understood and modeled in order to capture the insight required to improve the task, project, and industry level productivity.

At any given instance of the project, work can be measured by the area under the curve as Work $= \int Effort\ \partial(duration)$. The multidimensionality of the work can be reflected using a multi-layered model as is recognized in JPM, which allows the measurement to be extended across the multiple activities and trades and to take into account the effort and contributions at any given point.

FIG. 5 Work represented as a combination of non-linearity and dynamics of cost code weights over a project's duration.

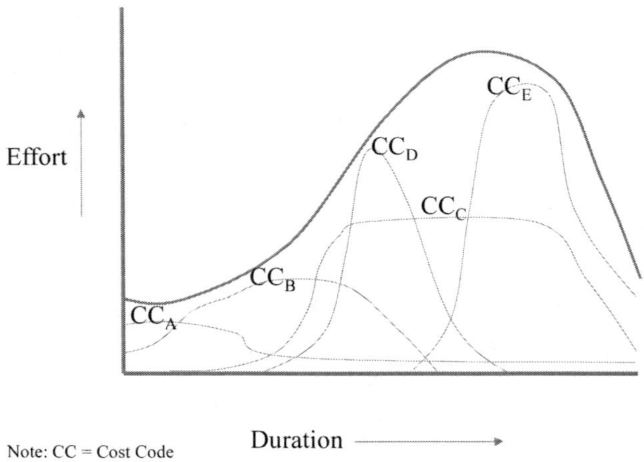

In addition to using the multi-dimensional view of work, measurements of productivity require true feedback about the work. This feedback will provide a unique set of information that can be expanded upon and used to reduce the interoperability that exists today in construction. Although building information modeling (BIM) increases the interoperability of information about the physical construction, the set of information built by use of JPM will focus on reducing the entropy of information[4] produced in the built environment. Information about jobsite progress is feedback more instantaneously based on daily activities contributing to completion, rather than waiting for the costs and/or time to report about progress after the fact.

ASTM E2691: Standard Practice For Job Productivity Measurement: Rationale and Principles

The rationale for developing the ASTM E2691 standard [1] was three-fold:

1. The method should be based on outcome, not output.
2. Construction productivity lags the national average. Although this assertion has been challenged by some research, the majority of national measurements point to the fact that construction productivity is behind other industries [5].
3. The measurement method should measure production rate change and allow for regular and instantaneous feedback to the stakeholders with early warning

[4]Information Entropy is a measure of *unpredictability* of *information content*.

signals of special causes impacting productivity on the jobsite so corrective action or learning can happen in a timely manner.

Building on these points, the principles that were used as the foundation to JPM are:

- Value transfer
- Definition and measurement of productivity
- Feedback from the source
- Statistical process control

VALUE TRANSFER

JPM encompasses all of the steps of industrialization, beginning with the management of labor and management of work. To improve productivity, other industries have had to understand how individual craftsmen perform their trade and recognize how that trade translates value to end users. This can only be captured with measures of outcome according to what the end user expects. Recognition of the activities, which enable transfer of value from designers, architects, engineers, builders, general contractors, subcontractors and distributors to the end user can only happen at the point of final work and installation. It is this level of recognition that allows for timely correction and learning. It is the deviation from expected outcome, which needs to be made visible, measured, and corrected in a timely manner. Any benchmarking that indiscriminately includes all the activities on the jobsite—both those that transfer value and those that result in rework and waste—will hide the true required resources to transfer the developed value to the final user.

Identifying which activities transfer the required value versus those which do not will help set more accurate benchmarks for the production rate measurement, thereby avoiding inclusion of hidden losses as part of the work performed on the job site. The activities on the jobsite can be grouped into three categories, value transferring, non-value transferring but necessary, and non-value transferring and not necessary. The true allocations of value transfer to the end user are such activities as:

- Planning
- Externalizing work (prefabrication)
- Quality
- Safety
- Reliability
- Predictability
- Final assembly
- Construction put in place
- Record keeping—as-builts, submittals, BIM
- Time and cost

The non-value transferring but necessary activities are such activities as:

- Inspection
- Moving material to the location

- Garbage removal
- Permits and government regulation compliance
- Labor limitation

The non-value transferring and not necessary activities are such activities as:

- Material handling
- Material movement (other than from delivery truck to the point of installation)
- Rework
- Repair
- Errors
- Unnecessary movement
- Following up on missing or incorrect information

Construction, like any other industry, has to not only rely on the skilled trades as the final installers but use their abilities to design the flow of value to the end user. The value perceived in the outcome of a project comes from the transfer of what all of the contributors to the project bring to the table, including conceptualization, design, engineering, planning, construction, operation, maintenance, and usage. Each of these players brings to the table their physical ability, their brainpower, and their collective talents to create a building or infrastructure project that will be used by the end customer and society. Waste in construction comes from not correctly and effectively transferring the skills and knowledge of the contributors into construction put in place. Any activity that happens on the project which contributes to construction-put-in-place transfers value; any other activities that does not (moving material, waiting on answers, rework) are waste. JPM is intended to first and foremost, make this effectiveness visible, and second to determine where the waste is occurring so it can be removed.

DEFINITION AND MEASUREMENT OF PRODUCTIVITY

As was discussed earlier, the measurement of productivity has been attempted over decades with several approaches. The definition of job productivity as explained in JPM is how effectively the incoming resources are transferred to project outcome. A measure that will match such a definition must take into account change in production rate, which JPM does. Measuring production only or production rate only will not provide the information about value transfer of resources. For example, **Fig. 6** shows measures of production, production rate, and change in production rate on two different days. On both days, the crew had 1000 ft (304.8 m) of production of pipe. The production rate on both days is also the same: 1000 ft (304.8 m) per 8 h, or 125 ft (38.1 m) per hour. The measure of change in production rate shows that the variation in production and production rate was much higher on day 2.

As a measure of change in production rate, JPM makes visible the variation in productivity during the job, plus deviation from expected outcome. Variation

FIG. 6 Scenario showing difference in measuring production, production rate, and change in production rate [24]

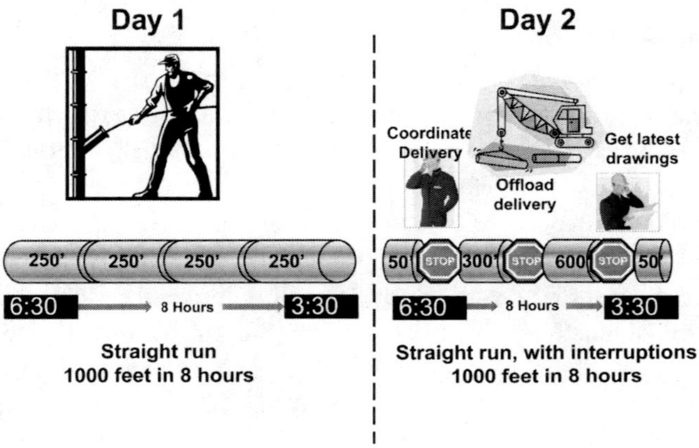

during the job can cause quality issues, safety concerns, and can contribute to increases in production rates due to the stop-and-start nature of the work. By measuring change in production rate, JPM will identify any non-production activities that would otherwise be included, or hidden, in the production measurements. For example, rework or repairs will show up as soon as it is happening because the production rate will decline.

In addition to the measurement of production rate and production rate change, JPM relies on the usage of outcome rather than just output of the work on jobsites. Production measures capture the quantitative output of work, such as lineal feet, cubic yards, number of fixtures set, etc. Productivity measures capture how effectively the input resources translate to final project outcome. To do this, the measurement of progress has to include more than quantitative output. JPM uses observed percentage complete of allowed work that translates to project completion as the outcome measurement. The allowed work that is measured takes into account installation ("production") and non-installation work as well as the difficulty of the work, which is only known by the technician. Moreover, when the labor is working on non-allowed work such as unapproved change orders or rework, JPM will show the decline in production rate because the labor hours being spent are not contributing to the project outcome. Non-contributing activities are hidden in measures of production output. For example, if one crew can rough in 5 rooms per day and another crew can rough in 10 rooms per day, the first crew may have a better production rate based on output. However, if over half of the rooms roughed in by the second crew have to be redone after inspection, then the first crew is in fact more productive.

JPM measures progress of the job relative to project outcome rather than just production output. JPM aggregates measures of task productivity to provide job productivity, and task productivity across projects or categories of projects can be aggregated to the industry level to provide a measurement of productivity that can be used as a national reference point by task and type of work [24].

FEEDBACK FROM THE SOURCE

Another principle that JPM follows is that, in construction, the installers are skilled operators and know how the building will be constructed. It is up to supervision and management on a project to understand what is getting in the worker's way from installing effectively. Taiichi Ohno [25], the father of the Toyota Production System, lived by this principle. He would request that every new hire stand and observe the production line for days to understand what was happening and to see what could be done to help improve the labor's time put toward effective production. The same must be done in construction projects. However, since there is no "field force" available and ready to stand on projects every day and observe this, JPM was designed to measure the construction work environment and make it as visible as if you were standing there every day watching. For example, **Fig. 7** shows a progression of events on a jobsite where a skilled tradesman tries to install a piece of conduit and is disrupted for lack of fuel in his lift.

FIG. 7 Jobsite scenario where labor is disrupted from installation.

Another example is shown in **Fig. 8** where the bends required in conduit due to concrete cores misaligned was four times what was expected. The crew may have done these bends very efficiently, but the effort they spent did not contribute to the outcome of the project as expected and productivity declined for that work.

Measuring this input requires data and information from the point of installation. Based on research from Moore (2013) [23], a conceptual model has been developed for construction information available at and from the point of installation. The information available at the point of installation includes any thoughts, decisions, and actions made where work occurs. This information is ever-present in the minds of the workers, and is therefore impossible to see unless they make the information explicit through writing or speaking. **Figure 9** shows a depiction of the concept for worker and crew actions and decision making when they encounter

FIG. 8 Jobsite scenario where installation effort deviated from expected outcome.

FIG. 9 Model of the jobsite information propagation in the specific scenario where workers encounter obstacles [23].

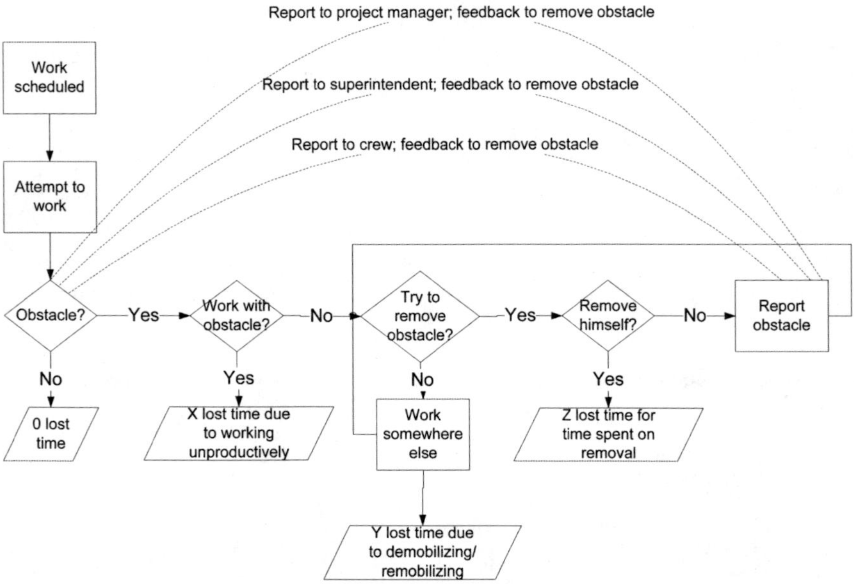

an obstacle while working. In the model, a worker has work scheduled (either by him/her or for him/her), and then attempts to work [23]. When an obstacle is encountered, the worker goes through a series of decisions about how to handle the obstacle. He/she either chooses to work around the obstacle, or to have the obstacle removed either by resolving it himself/herself or reporting it and getting help elsewhere.

This scenario involves several pieces of lost information, if the obstacle is not reported beyond the worker or the crew level, including:

- What the obstacle was and what caused it?
- How the worker resolved the obstacle?
- What was the impact of the obstacle?
- What is the likelihood of obstacle reoccurrence?

Regardless of the obstacle handling choice, there is a loss of time. The lost time has two impacts: (1) workers lose effective time on installation, and (2) the scheduled work takes longer than expected to complete. One or both losses could occur. The result of both is additional cost to the project and at a macro level, to the industry, potentially without any signal of productivity loss.

Modeling approaches used in other industries that are characterized as complex production were used to model and explore this information and feedback. A combination of agent based modeling and social network analysis showed that 36 % of the time, workers resolve the obstacles they encounter on their own; 40 % of the time the workers wait for help, and they work around the obstacle without any resolution 24 % of the time. In addition, although workers encounter obstacles in almost nine out of ten scheduled work tasks, they report the obstacle beyond their crew 50 % of the time. The research also showed that information is generated based on a combination of the workers' prior knowledge and experience, and the local conditions at their point of installation.

The information that the workers have about their own experiences on jobsites, as well as their technical trade and skill is kept to themselves and used by themselves to locally adapt and coordinate in their daily work. This information becomes explicit particularly in the face of obstacles, when they have to improvise and decide how to handle the obstacles, either through resolving on their own, waiting for help, or working around the obstacle. As the obstacle is being handled or following its handling, information about the obstacle or incident is sometimes passed on beyond the crew. If passed outside the crew, it can help others on the jobsite to resolve future obstacles. If the information is not passed on outside the crew that faced the obstacle, the only other means for information passing is through the social network of the workers onsite. In this case, the information is still not carried out beyond the workforce or the jobsite, and therefore cannot be understood to solve work performance problems onsite.

The conceptual model and data collection and analysis used to populate the model showed that information is more likely to propagate through the workers onsite than through the formal hierarchy of the jobsite or crew structure. This means that without an appropriate measurement method relying on input from workers, how value transfers or does not transfer in construction is completely hidden. The output and even outcome of the transfer will be known in time, but what causes it to transfer effectively and efficiently would not be known otherwise.

This model and research [23] verifies the hypothesis of JPM that information and feedback from the source at the jobsite is necessary for correct and accurate productivity measurement.

STATISTICAL PROCESS CONTROL

JPM utilizes principles of statistical process control for managing variation and using the measurement and signals for feedback and improvement. The science developed for manufacturing by Dr. Shewhart has been applied to the construction environment for recognition of special causes and common causes of variation in productivity differential [26]. The measurement of deviation from expected outcome can be utilized at the task, project, and industry level for measuring variation within and between projects, types of work, geographical sectors, project delivery methods, and many other factors.

Application of ASTM E2691

As the first model of productivity that includes information from the source of work in construction, JPM measures construction productivity at three levels still relying on the principles explained earlier in this paper: task, project, and industry. **Figure 10** illustrates the three levels using the definitions below [1]:

TASKS

Tasks refer to specific construction activities such as concrete placement or structural steel erection.

PROJECTS

Projects are the collections of tasks required for the construction of a new facility or structure (e.g., the construction of a new commercial office building or highway infrastructure) or renovation (i.e., additions, alterations, and major replacements) of an existing constructed facility

INDUSTRY

Industry represents measures that are based on the North American Industrial Classification System (NAICS) codes for the construction sector and would represent the total portfolio of projects.

By comparing labor hours used against construction-put-in-place (CPIP), JPM allows for unified measurement of established building elements according to the UNIFORMAT II format [27]. ASTM E2691 starts with the background for and

FIG. 10 Concept of application of JPM at task, project, and industry levels [2].

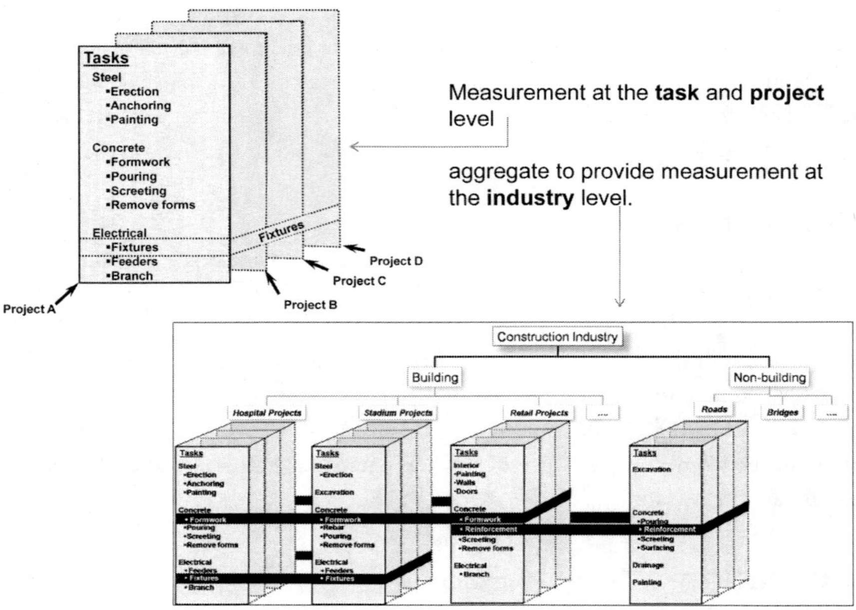

significance of the development and follows with the explanation of the mechanics of the practice.

JPM measures construction productivity differential on both an ongoing and a periodic basis, as well as average productivity over the life of the construction project. JPM also measures ongoing changes in the production rate.

JPM calculates the ratio of outcome per unit of input: how much work—Construction-Put-In-Place (CPIP)—was produced by how many labor hours. The data sources for the JPM, which will be explained in the following sections, require input from the source of CPIP, which is the installer himself. He or she is the only one who knows that the CPIP was complete, correct, and contributing toward final installation. To accomplish this, the measurement must be used to gather information from the source and used to correct and improve system productivity, not individual productivity. The results of the JPM need to be shared with the installer, site supervision, managers, owners, and local and national authorities. The standard practice requires this information and feedback in order for the measurement to be useful.

Additionally, JPM is an early warning signal for construction performance. It measures ongoing productivity changes, trends, and anomalies resulting from changes on a construction jobsite, which enables contractors, project managers, supervisors, and foremen to react and improve productivity as the construction project unfolds.

The JPM standard practice establishes a process for measuring construction job productivity by comparing labor usage to CPIP.

- JPM measures labor productivity of the installation processes on a construction job.
- CPIP is measured with input from the labor performing the installation, utilizing elements of statistical process control (SPC) and industrial engineering.
- JPM takes into account the difficulty of installation at any given point on a job.
- JPM evaluates relative productivity changes using trend monitoring.

Overall JPM's informative output can be summarized in the following points

- **JPM produces two measurements: construction production rate and productivity (in other words, the change of production rate).**
  - o JPM measures the overall production rate by comparing observed on hand CPIP to the expected production rate.
  - o JPM measures overall job productivity through a comparison of labor usage to a labor productivity reference point.
- **JPM issues early warning signals for construction productivity changes and anomalies**
  - o JPM identifies productivity deviations in the form of any gains or losses in productivity, and anomalies indicating a special cause, from the productivity reference point.
  - o JPM measures the productivity changes to individual elements of the project so that the causes of productivity trends can be further investigated. These elements can be an industry accepted code of accounting or business

specific cost codes if the company using JPM already has them in place. They can also be individual building elements according to the UNIFORMAT II format for organizing building data, in ASTM E1557 [27]. The use of this classification will allow for task, project, and industry level comparison of common elements. JPM measures ongoing changes in labor usage. It is not a one-time measurement that is done at a single point in the project.

- **JPM measures productivity and its changes wherever the labor or any human resources are used in a construction project by:**
  - o Any contractor or construction manager directly or indirectly responsible for the productivity of the human resources and its usage.
  - o Any contractor or construction manager conducting self performance on any portion of the construction job.
  - o Any contractor or construction manager supervising labor and human resources performance on any portion of a construction job.
  - o Any entity performing any kind of intellectual work such as planning, scheduling, prefabrication, layout, building modeling, material management, safety, site management, trade and task coordination or any kind of non-installation activities either required or contributing to the project's progress.
  - o Any required regulatory activities such as inspections, commissioning, sustainability requirements and LEED requirements.

In addition to using JPM during a project, the feedback from the measurement and approach can be fed back to estimating for improved accuracy during the bidding process. The information gathered from JPM is an enhancement to existing estimating and accounting databases which rely on historical averages and after-the-fact measurement. JPM provides an operational database that captures the information from the point of installation as the work unfolds. This data can be collected and mined with many factors such as type of work, geography, crew mix, site conditions, and cost code mix of the work to provide a comparison of estimated hours, baseline labor hour budget (BLHB), and accounting hours that shows the average and range of deviations from expected performance. Beyond company level, local, regional, and national authorities can use the data and analysis to understand performance differences and develop systematic approaches to improvement in competitiveness and productivity.

Usage of JPM is relatively simple. The standard explains how to create the baseline labor hour budget using work breakdown structure, how to measure the productivity throughout the job, and how to read and interpret the outputs. The high-level steps of using JPM include:

- Establish a BLHB based on the historical labor units used to perform any specific task, installer's input, or any industrial acceptable benchmarks for the scope of the construction job being measured using a work breakdown structure (WBS) and reference to the UNIFORMAT II Classification (ASTM E1557) [27].
- Evaluate the BLHB for appropriate level of detail.
- Establish the labor productivity reference point (LPRP).

- Once any labor hours are expended on the job (even before installation commences, with activities such as planning, layout, pre-assembly), begin tracking the JPM.
- Report the JPM productivity differential (PD) and review the results for signals of special causes impacting the productivity.

Once the LPRP is established, the JPM is calculated on an ongoing and regular basis to measure productivity on the project. The frequency of the updates is a function of the project speed. Typically a once a week measurement of the job productivity and observed % completion is sufficient for most projects. However, if the job is a fast track project and requires more labor, the frequency can increase to a few times a week or even daily. The JPM is calculated based on the regular collection of CPIP (through observed % complete measurement) and comparison of it to the expended labor hours versus the LPRP.

The reporting in JPM relies strongly on the principles of SPC. The science of SPC was developed to help manage and make sense of large quantities of data required to establish the change in the rate of production. Additionally, SPC enables the user to distinguish between common[5] and special[6] causes as well as recognize when a common cause should be considered and dealt with as a special cause. This application is relevant at the task, project, and industry level; however, the basis for the data collection for JPM comes from the tasks on individual jobsites and is aggregated to higher levels by weighting the tasks and jobs using the same approach as shown in **Fig. 5**.

The PD results are also trended graphically over time, with the 0 % line representing the LPRP. Each PD measurement is plotted shown as a trend over time by job and by cost code. When the productivity differential is above the line, interpret that the job productivity is better than planned according to the initial LPRP (**Fig. 11**). When the productivity differential is below the line, interpret that the job productivity is worse than planned according to the initial LPRP.

TASK, PROJECT, AND INDUSTRY JPM APPLICATION

JPM can be used for measuring productivity at the task, project, and industry level. **Figure 12** shows an example of reporting and trending productivity differential on a *project* and in four different activities on one project for an electrical contractor. This reporting allows a single contractor to see how task productivity is contributing to the contractors' overall project productivity. At this level of application, JPM is used to monitor job productivity and evaluate deviations during a project to catch special causes of variation in productivity that could impact overall job outcome.

In addition to a single contractor's review of productivity, JPM provides a method for measuring overall *project* productivity, in which all trades can be

---

[5]Common Cause is a source of variation that is normal and expected, common cause variation is predictable and cannot be traced back to a single source.

[6]Special Cause is a source variation that causes fundamental change in a process. Special cause variation signals a change in a process and can be traced back to a source.

FIG. 11 Interpretation for JPM graph reporting [1].

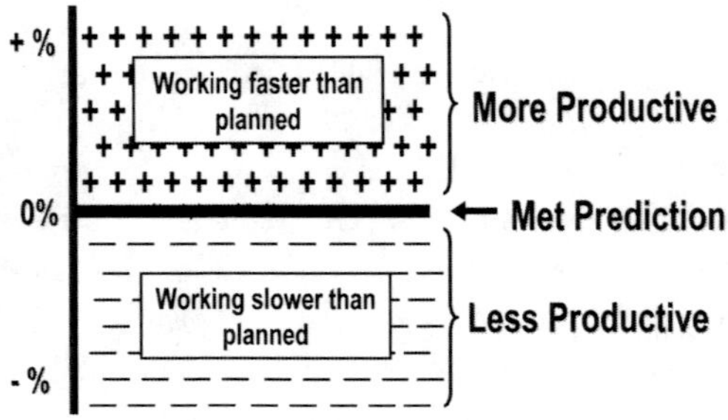

aggregated with an overall view (see **Fig. 13**). In this application, the construction manager or general contractor uses the same approach as shown in **Fig. 4** for monitoring the trend of each subcontractor's PD. The "total job" PD shown in **Fig. 12** would be plotted as one sub's productivity differential in **Fig. 13**. These trends can be evaluated individually, and also weighted and rolled up to provide an overall job PD that relies on a consistent JPM method for measurement.

Finally, the ASTM E2691 standard can be used to aggregate productivity at an *industry* level (see **Fig. 14**). In this application, a standard set of tasks that apply across different industry sectors can be measured and monitored. Various analyses can be conducted to understand differences in productivity between and within regions, types of work, project delivery methods, and many other factors. The JPM method provides a common approach that is "bottom up" and proven to correctly measure productivity according to the above-mentioned principles. As such, the measurement outcome can be consistently used by an authority to understand the

FIG. 12 Example reporting at project level for an electrical contractor.

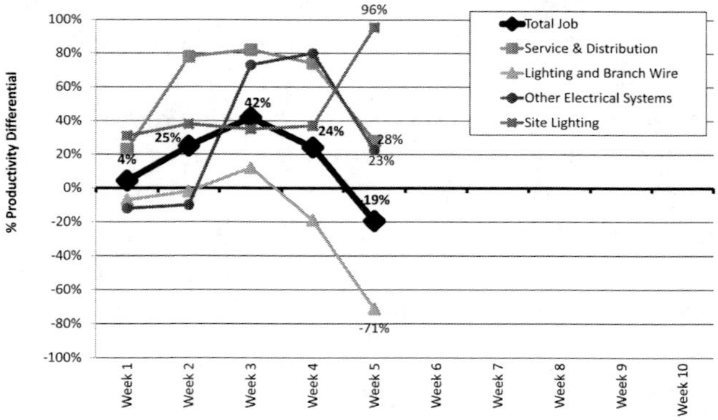

FIG. 13 Example reporting at project level for all contractors

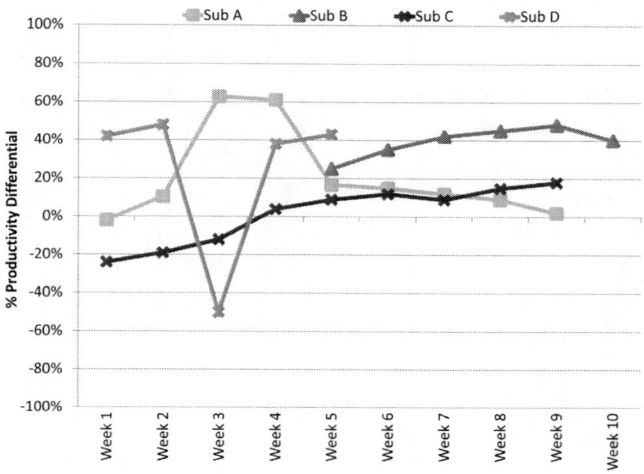

industry's productivity benchmarks and trends. To do so, an industry standard classification would need to be used for common task level definition.

Individual contractors or stakeholders can also use the information collected with JPM as input to their estimating, accounting, and scheduling processes. The input of real-time and instantaneous feedback can be valuable inputs to other databases used in overall project management.

FIG. 14 Example reporting for a task across sectors of the construction industry.

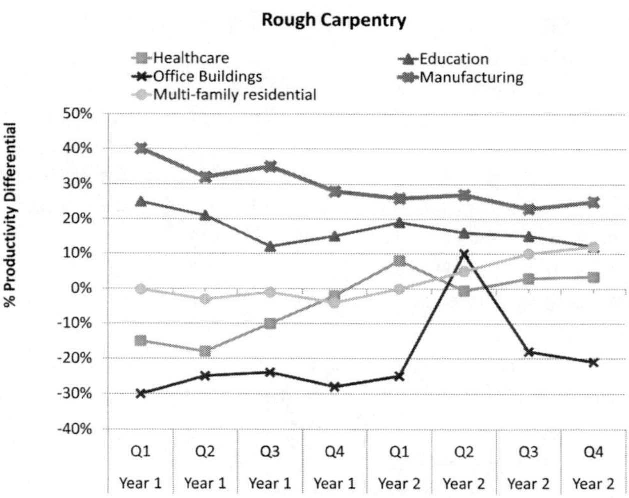

Conclusion

ASTM E2691 [1] provides a method for measuring task, project, and industry level productivity. Given the findings of the National Research Council [6] and NIST [5], ASTM Subcommittee E06.81 developed this standard to help improve construction productivity to match what other industries that have went through industrialization. This industry can benefit in many ways by using the standard for measuring productivity and having feedback from the source of information in construction, which today is skilled trade. This information channel needs to be further understood, modeled, and explored to translate the knowledge of centuries of physical work into physical output that can multiply today's output. Consumers of construction applying this standard should be able to see higher quality, lower cost, and faster outcomes for their expectations.

Appendix

The nomenclature below is referenced within the body of the paper, and further defined or explained:

1. Agent based modeling (ABM): a modeling approach used for simulating the actions of autonomous and interactive agents, by programming simple results that often result in an emergent system-level behavior that is independent of any single agent or combination of agents.
2. Baseline labor hour budget (BLHB): a budget of direct labor hours created at the onset of a new construction project that approximates how many hours will be spent on any defined part of the project. The budgeted hours are first assigned to the tasks on the project, and can be summed to determine budgeted hours for any cost code or for the entire project.
3. Construction put in place (CPIP): a measure tracked by the U.S. Census Bureau representing the dollar value of construction work done, as paid for by the end-user of the construction.
4. Cost code (CC): a categorization of activities or common physical elements of a project that are tracked typically in construction estimating and accounting systems.
5. Labor productivity reference point (LPRP): a ratio calculated at the beginning of a construction project, for the hours needed to complete one % of the construction, based on the baseline labor hour budget.
6. Productivity differential (PD): a measurement of the % difference between the labor productivity reference point and the current labor productivity for the given timeframe.
7. Social network analysis (SNA): method for studying networks representing social connections of actors in a network, typically involving topographical maps and mathematical models that can be used to model actor interactions and their impact within and between networks.
8. Work breakdown structure (WBS): the outcome of a process that decomposes a project into layers that connect the top level project deliverable to the bottom

level finite tasks representing the work that needs to be done to complete the project deliverables.

References

[1] ASTM E2691-11: Standard Practice for Job Productivity Measurement, *Annual Book of ASTM Standards*, ASTM International, West Conshohocken, PA, 2013.

[2] ASTM MNL 65, *Manual on Standard Practice for Job Productivity Measurement*, ASTM International, West Conshohocken, PA, 2012.

[3] Bureau of Economic Analysis, 2011, "Gross-Domestic-Product-(GDP)-by-Industry Data," Bureau of Economic Analysis, http://bea.gov/industry/gdpbyind_data.htm (Last accessed 12 June 2011).

[4] Central Intelligence Agency, 2009, "The World Factbook 2009," https://www.cia.gov/library/publications/the-world-factbook/geos/xx.html (Last accessed 12 Dec 2009).

[5] Huang, A., Chapman, B. and Butry, D., *Metrics and Tools for Measuring Construction Productivity: Technical and Empirical Considerations*, National Institute of Standards and Technology, Gaithersburg, MD, 2009.

[6] The National Academy of Science, *Advancing the Competitiveness* and *Efficiency of the U.S. Construction Industry*, Board on Infrastructure and the Constructed Environment, Division on Engineering and Physical Sciences, National Research Council, Washington, D.C., 2009.

[7] Daneshgari, P. and Moore, H., *Industrialization of the Construction Industry*, IEC Insights, Alexandria, VA, 2013.

[8] Haas, C. T., Borcherding, J. D., Allmon, E. and Goodrium, P. M., "U.S. Construction Labor Productivity Trends 1970–1998," *J. Constr. Eng. Manage.*, Vol. 126, 2000, pp. 97–104.

[9] Oglesby, C. H., Parker, H. W. and Howell, G., *Productivity Improvement in Construction*, McGraw-Hill, New York, 1989.

[10] Park, H., Thomas, S. R. and Tucker, R. L., "Benchmarking of Construction Productivity," *J. Constr. Eng. Manage.*, Vol. 135, 2005, pp. 772–778.

[11] Gilbreth, F. B. and Gilbreth, L. M., *Fatigue Study: The Elimination Of Humanity's Greatest Unnecessary Waste; A First Step In Motion Study*, Macmillan, New York, 1916.

[12] Adrian, J. and Boyer, L., "Modeling Method Productivity," *J. Constr. Div.*, Vol. 102, 1976, pp. 152–168.

[13] Thomas, R. and Yiakoumis, I., "Factor Model of Construction Productivity," *J. Constr. Eng. Manage.*, Vol. 113, 1982, pp. 623–639.

[14] Dai, J., Goodrum, P. M. and Maloney, W. F., "Construction Craft Workers' Perceptions of the Factors Affecting Their Productivity," *J. Constr. Eng. Manage.*, Vol. 135, 2009, pp. 217–226.

[15] Noor, I., "Measuring Construction Labor Productivity by Daily Visits," *AACE International. Transactions of the Annual Meeting*, Cincinnati, OH, AACE Inc., Morgantown, WV, 1998.

[16] Thomas, H. R., Sanders, S. R. and Bilal, S., "Comparison of Labor Productivity," *J. Constr. Eng. Manage.*, Vol. 118, 1992, pp. 635–650.

[17] Sawhney, A., Bashford, H., Walsh, K. and Mulky, A. R., "Agent-Based Modeling and Simulation in Construction," *Proceedings of the 2003 Winter Simulation Conference, IEEE*, Dec. 3, 2003, pp. 1541–1547.

[18] Macal, C. and North, M., "Tutorial on Agent-Based Modeling and Simulation," *Proceedings of the 2005 Winter Simulation Conference, IEEE*, M. E. Kuhl, N. M. Steiger, F. B. Armstrong, and J. A. Joines, Eds., Orlando, FL, Dec 4–7, 2005, pp. 2–15.

[19] Desai, A., 2012, "Exploring Crew Behavior During Unexpected Events in Construction," M.S. thesis, Michigan State University, East Lansing, MI.

[20] Watkins, M., Mukherjee, A., Onder, N. and Mattila, K., "Using Agent-Based Modeling to Study Construction Labor Productivity as an Emergent Property of Individual and Crew Interactions," *J. Constr. Eng. Manage.*, Vol. 135, 2009, pp. 657–667.

[21] Choy, E. and Ruwanpura, J., "Situation Based Modeling for Construction Productivity," *Can. J. Civ. Eng.*, Vol. 33, 2006, pp. 1585–1600.

[22] Ballard, G., Tommelein, I., Koskela, L. and Howell, G., "Lean Construction Tools and Techniques," *Design and Construction: Building in Value*, R. Best and G. de Valence, Eds., Butterworth-Heinemann, Woburn, MA, 2002, pp. 229–233.

[23] Moore, H., 2013, "Exploring Information Generation and Propagation from the Point of Installation on Construction Jobsites: An SNA/ABM Hybrid Approach," Ph.D. thesis, Michigan State University, East Lansing, MI.

[24] Daneshgari, P., *Agile Construction for Electrical Contractors*, Jones & Bartlett, MA, 2009.

[25] Ohno, T., *Worklpace Management*, Gemba Press, Mukilteo, WA, 2007.

[26] Shewhart, W. A., *Economic Control of Quality of Manufactured Product*, American Society for Quality Control, Milwaukee, WI, 1931.

[27] ASTM E1557-09: Standard Classification for Building Elements and Related Sitework-UNIFORMAT II, *Annual Book of ASTM Standards*, ASTM International, West Conshohocken, PA, 2011.

STP 1586, 2014 / available online at www.astm.org / doi: 10.1520/STP158620140035

Douglas S. Thomas[1]

Economic Impact of Improved Service-Life Prediction for Seams in Low-Slope EPDM Roofing

Reference

Thomas, Douglas S., "Economic Impact of Improved Service-Life Prediction for Seams in Low-Slope EPDM Roofing," *Ideas to Impact: How Building Economic Standards Keep You on Track*, STP 1586, Robert E. Chapman and Michael N. Goodkind, Eds., pp. 127–148, doi:10.1520/STP158620140035, ASTM International, West Conshohocken, PA 2014.[2]

ABSTRACT

In the early part of the 1970s, built-up roofing dominated the U.S. membrane roofing industry. It was estimated that 10 % to 15 % of these roofs failed prematurely, but by the early 1980s new membranes were being rapidly adopted. Among them, ethylene-propylene-diene terpolymer (EPDM) rubber, manufactured as preformed single-ply sheets ready for field installation, experienced the most rapid growth. Although performance was generally satisfactory, far too often, disbonding of seams occurred. National Institute of Standards and Technology (NIST) researchers developed a creep-rupture methodology to study the factors affecting seam performance including mechanical loading of the seam, thickness of the adhesive used to form the seam, and rubber surface cleanness of the rubber before the adhesive was applied. This research had a noteworthy impact on the EPDM roofing industry in that it accelerated changes to seam technology and increased users' confidence that EPDM would provide satisfactory long-term performance. This paper examines the economic impact of the research conducted by NIST's Service Life Prediction (SLP) Program for EPDM roofing. Three ASTM standards are used to calculate the value of NIST's research investment: (1) present value of net savings (PVNS) [ASTM E1074-09: Practice for Measuring Net Benefits and Net Savings for Investments in Buildings and Building Systems, *Annual Book of ASTM Standards*, ASTM International, West

Manuscript received April 1, 2014; accepted for publication May 8, 2014; published online July 14, 2014.
[1]Economist, Applied Economics Office, Engineering Laboratory, National Institute of Standards and Technology, Gaithersburg, MD 20899, United States of America, e-mail: douglas.thomas@nist.gov
[2]ASTM Symposium on *Ideas to Impact: How Building Economic Standards Keep You on Track* on October 5, 2014 in New Orleans, LA.

Conshohocken, PA, 2012], (2) savings-to-investment ratio (SIR) [ASTM E964-06: Practice for Measuring Benefit-to-Cost and Savings-to-Investment Ratios for Buildings and Building Systems, *Annual Book of ASTM Standards*, ASTM International, West Conshohocken, PA, 2012], and (3) adjusted internal rate of return (AIRR) [ASTM E1057-06: Standard Practice for Measuring Internal Rate of Return and Adjusted Internal Rate of Return for Investments in Buildings and Building Systems, *Annual Book of ASTM Standards*, ASTM International, West Conshohocken, PA, 2012]. The results show that the PVNS attributable to NIST is $\$26.7 \times 10^6$ with an SIR of 2.5 and an AIRR of 13.7 %. Stated in present-value terms, every public dollar invested in NIST's SLP-related research, development, and deployment effort generated $2.5 in cost savings to the public.

Keywords

adjusted internal rate of return, construction, EPDM, present value net savings, roofing, savings-to-investment ratio, service-life

Introduction

In 1969, it was estimated that approximately 10 % to 15 % of built-up roofs (BUR) failed prematurely [1]. Although there were prescriptive specifications indicating the type and number of reinforcing plies and the type and amount of bitumen, specifications detailing the performance requirements for completed BUR membranes did not exist. A common result was that installed membranes had inadequate properties to perform satisfactorily. In 1974, however, the National Institute of Standards and Technology (NIST) researchers Robert Mathey and William Cullen published Preliminary Performance Criteria for Bituminous Membrane Roofing [2]. The performance concept that was applied to BUR membranes was widely accepted by industry and membranes were beginning to be selected by conformance to the criteria. Roofing contractors were among the most supportive, as the National Roofing Contractors Association (NRCA) incorporated performance properties into its manual.

At the beginning of the 1970s, built-up roofing had a monopoly on the U.S. membrane market. By the early 1980s, however, new membranes were being rapidly adopted and ethylene-propylene-diene terpolymer (EPDM) rubber, manufactured as preformed single-ply sheets ready for field installation, experienced the most rapid growth. Although performance was generally satisfactory, far too often, disbonding of seams occurred.

By the mid-1980s, EPDM roofing accounted for about 35 % of the membrane market. EPDM is a rather chemically inert rubber, which makes it attractive for outdoor use as a membrane material. However, this chemical inertness becomes a limitation when bonding adjacent sheets in the field to form the seams of a waterproofing membrane. At the time, these seams were typically fabricated with contact-type, polymer-based, liquid adhesives. In the mid-1980s, unsatisfactory seam performance accounted for about 50 % of the EPDM membrane problems reported to the NRCA in surveys of member contractors. In 1994, NIST initiated an industry–government consortium of roofing stakeholders to (1) compare the performance of tape-bonded and liquid-adhesive-bonded seams of EPDM membranes used in roofing, and (2) recommend a test protocol and criteria for evaluating the performance of such seams. Participants included EPDM manufacturers, tape-system manufacturers, trade associations, and roofing consultants, designers, and appliers. Members of this consortium team were asked in 2009 what benefits they felt were brought about by the improved service life prediction that resulted from consortium activities. They responded with multiple benefits: decreased costs from substituting fewer or less-expensive materials (less over engineering); increased confidence of installers, designers, and customers that the new system would work and last; reduced repair costs from callbacks to fix leaks; decreased warranty repair costs arising from better predictability of lifetime system performance; and longer warranty periods—for example, 10- to 15-year warranties being replaced with 25- to 35-year warranties, which was later examined as part of an economic impact study. A NIST publication was developed [3] to examine NIST's investment in improving the service life prediction of polymeric materials, including EPDM roofing. One of the benefits that it examined is the benefit of warranty repair costs. This paper builds upon the work done in that report.

In 2007, EPDM was the single largest roofing material type for low-slope roofs—26.9 % of new construction and 24.3 % of reroofing [4]. In 2008, there was $\$3.4 \times 10^9$ in low-slope EPDM roofing for new construction and an additional $\$1.2 \times 10^9$ for reroofing, resulting in $\$4.7 \times 10^9$ in expenditures. Low-slope roofing accounted for $\$17.9 \times 10^9$ of the total $\$26.8 \times 10^9$ in new roofing and reroofing occurring in 2008 with the remaining $9.0 billion being steep-slope roofing associated with residential roofing.

This paper examines one of the benefits that were brought about by improved service life prediction—the reduction in warranty service repair costs in EPDM roofing. Although it is only one of the expected benefits from better-performing EPDM roofing systems, it is a significant benefit that has been documented. So while not all benefits or savings from improved EPDM systems are calculated, we can provide a lower-bound estimate of the savings based on actual records of warranty service costs. NIST technical note 1650 [3] examines the economic impact of research conducted through its Service Life Prediction Program for high-performance polymeric construction materials, which includes an examination of (1) coatings, (2) sealants, and adhesives, and (3) roofing. This paper expands upon the economic impact assessment of NIST's work on EPDM roofing. Three ASTM standards are used to calculate the value of NIST's research investment: (1) present value of net savings (PVNS) (ASTM E1074 [5]), (2) savings-to-investment ratio (SIR) (ASTM E964 [6]), and (3) adjusted internal rate of return (AIRR) (ASTM E1057 [7]). Variations in the input variables for these measures make it necessary to examine the uncertainty in the PVNS, SIR, and AIRR. The sensitivity analysis employed in this paper is based on techniques for treating uncertainty and risk described in ASTM E1369-11 [8].

Data

The data for this analysis are from EPDM Roof System Performance: An Update of Historical Warranty Service Costs by James L. Hoff [9]. The database underlying this study covers over 278.7×10^6 m^2 (3×10^9 ft^2) of EPDM roofs on over 150 000 roofs installed between 1982 and 2003. This study assumes that the performance of the EPDM roofs installed between 1982 and 1993 documented in this sample is representative of all those EPDM roofs installed during this time period. This is the best performance estimate that can be made given the data limitations. Each roof in the database has a material and workmanship warranty, the majority being 10 years in length. The date, cost, and type of service repairs are given for the entire warranty period. As of 2003, there were service repair data available for at least 10 years on all roofs installed in 1993 or before.

Because the historical warranty costs are exclusively the manufacturer's service call costs, these data might underestimate total repair costs; that is, owners sometimes incur additional repair/maintenance costs beyond what is covered in the warranty. For comparison purposes, the costs are presented in \$/m^2 (\$/ft^2). To preserve confidentiality of manufacturers' actual costs, the service costs described in Hoff's study are normalized and indexed to a baseline year, in this case 1987, with a cost index of 10.76 cost/m^2 (1.00 cost/ft^2). The 10-year stream of indexed warranty service cost data for any given installation year were adjusted for the time value of money using the U.S. National GDP deflator as calculated by the U.S. Bureau of Labor Statistics.

Figure 1 shows, as of July 2003, the change in total warranty service costs, by installation year, from 1982 to 1993, where the height of each bar in the figure represents, for that year's installed EPDM roofs, the cumulative, time-adjusted, indexed dollar cost per m^2 (ft^2) of the 10 years of warranty service. From 1982 to 1987, unit warranty service costs dropped from 70.0 to 10.8, an 85 % decline. And although the curve flattens out from 1987 to 1993, the warranty service costs still drop from 10.8 to 4.3, an additional 60 % decrease. From such significant drops in warranty costs from 1982 to 1987, Hoff [9] suggests that technological changes must have occurred. This could have taken place in roof materials, application technique, or design.

Hoff's article provides further insight as to just what component of the roofing was responsible for most of the service costs. **Figure 2** shows the indexed dollar cost per m^2 (ft^2) of warranty service costs by key components of EPDM roofing. Field seams were by far the most significant cost driver, and followed by perimeter flashings. Each of these decreased in 10-year unit costs between 1982 and 1987—87 % for seams and 90 % for flashings. Hoff's article provides a final insight that enables us to develop estimates of the reductions in warranty service costs. **Figure 3** shows the first 10-year warranty service costs for 1985 through 1993 as a percentage of the original installed costs.

Table 1 shows the estimated installed costs for EPDM roofs over the 9-year period, 1985 through 1993, reported in Hoff's article. These data, once linked with

FIG. 1 EPDM warranty service costs: First 10 years of service: (*a*) indexed cost per m^2, and (*b*) indexed cost per ft^2.

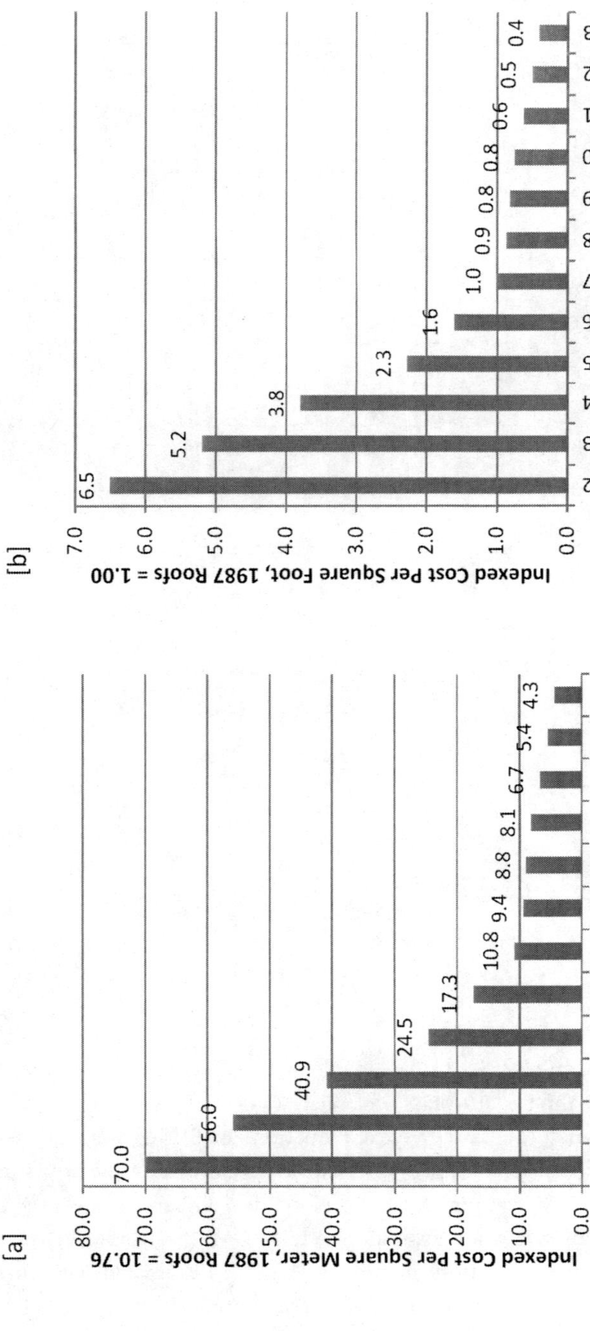

FIG. 2 EPDM warranty service costs by key component: first 10 years of service: (*a*) indexed cost per m^2, and (*b*) indexed cost per ft^2.

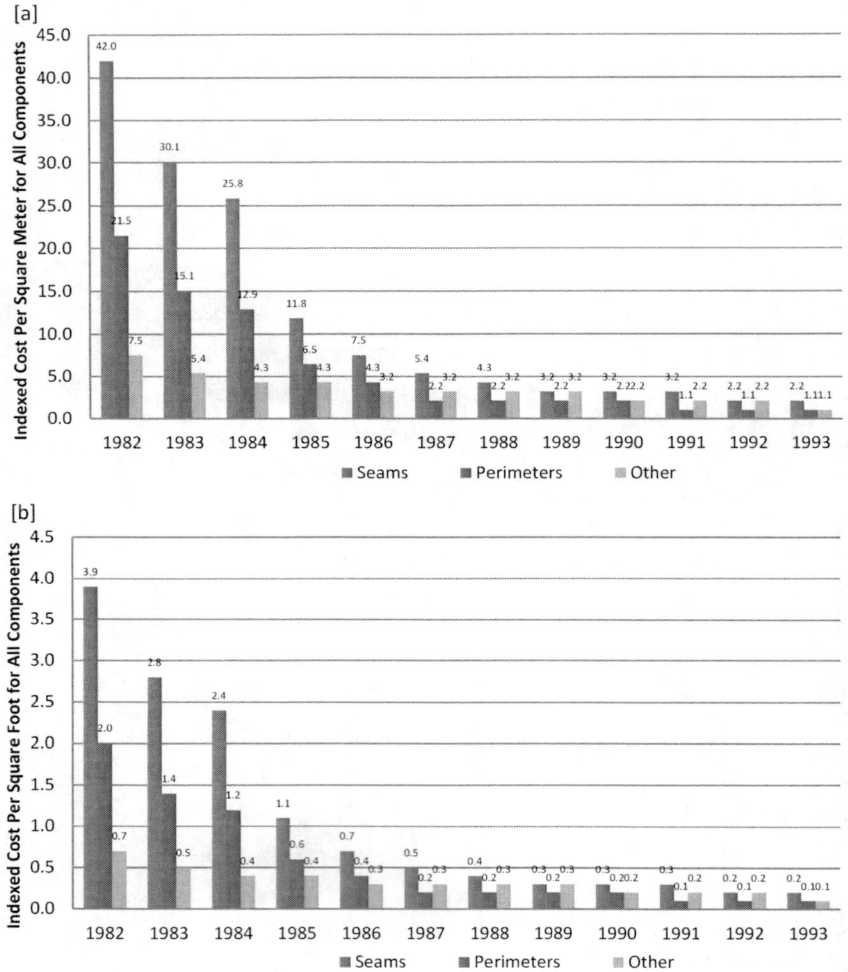

information contained in **Figs. 1**, **2**, and **3**, enable us to estimate EPDM-related warranty service costs for the first 10 years of service. These costs can then be annualized and used to estimate the year-by-year reductions in EPDM warranty service costs because of technological improvements. **Table 2** provides the investments made by NIST in EPDM roofing. NIST costs are for all service life prediction investments and do not break out costs specific to EPDM; therefore, EPDM's proportion of the service life prediction investments has been estimated. This proportion is shown in column 3 of **Table 2**. EPDM is a significant portion of the total in the early years, reducing in steps to a very modest amount in later years.

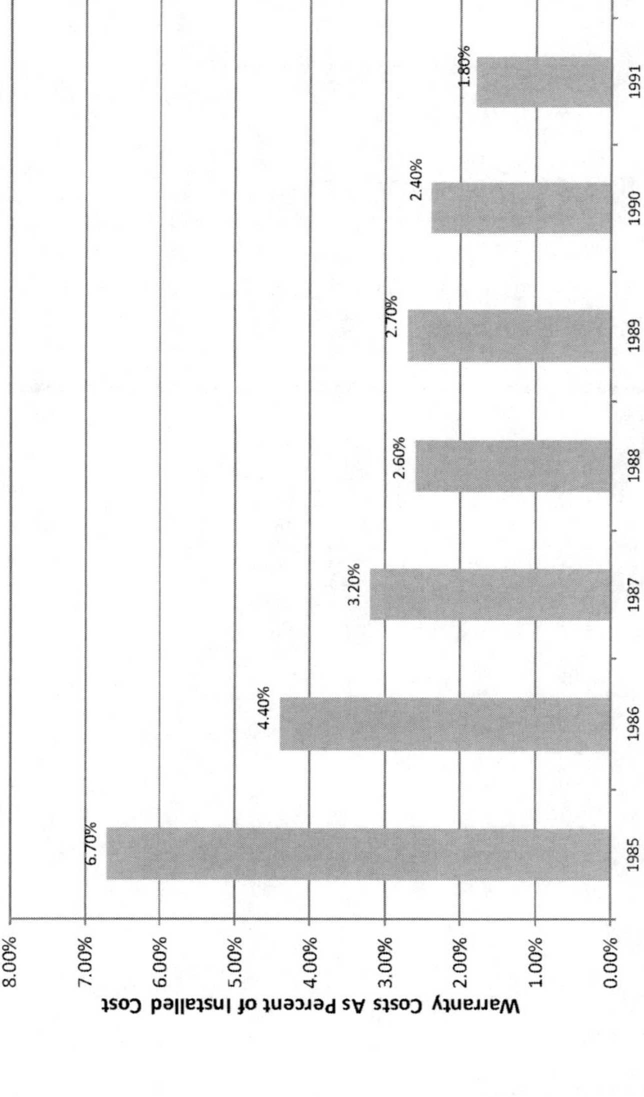

FIG. 3 EPDM warranty service costs as a percentage of original installed cost: First 10 years of service.

TABLE 1 Estimated installed costs for EPDM roofs: 1985–1993 [13–17].

Year	Installed Costs in Millions of Dollars
1985	1554.7
1986	1688.3
1987	1836.2
1988	1931.7
1989	1898.0
1990	2097.8
1991	2094.7
1992	2143.0
1993	2330.8

Methods

ASTM INTERNATIONAL STANDARD PRACTICES

Several methods of economic evaluation are available to measure the economic performance of a research program, a new technology, a building, a building system, or like investment, over a specified time period. These methods include, but are not limited to, present value of net benefits, present value of net savings, benefit-to-cost

TABLE 2 NIST research investments in EPDM roofing: 1994–2008.

Year	Present Value of Total SLP Investment Costs by Year (in Millions of 2008 Dollars)	EPDM Roofing Share	Present Value of EPDM Investment Costs by Year (in Millions of 2008 Dollars)
Column 1	Column 2	Column 3	Column 4 (2 × 3)
1994	1.705	0.950	1.620
1995	2.589	0.950	2.460
1996	2.947	0.950	2.800
1997	2.815	0.950	2.674
1998	2.388	0.600	1.433
1999	2.486	0.600	1.492
2000	2.322	0.600	1.393
2001	2.795	0.250	0.699
2002	3.338	0.250	0.834
2003	3.402	0.250	0.850
2004	2.386	0.250	0.597
2005	2.870	0.250	0.717
2006	2.345	0.050	0.117
2007	2.257	0.050	0.113
2008	1.901	0.050	0.095
Total			17.894

ratio, savings-to-investment ratio, and the adjusted internal rate of return. These methods differ in the way in which they are calculated, and, to some extent, in their applicability to particular types of investment decisions. The five methods described in this section are based on ASTM International standard practices [10]. Detailed descriptions of each of the standardized methods are given in Chapman and Fuller [11]. Readers interested in an in-depth survey covering these, as well as other methods are referred to Ruegg and Marshall [12].

To describe each of the five standardized methods, it is necessary to first introduce and define a series of terms. These terms are used to define each of the standardized methods. Throughout this section, the following terms are used as the basis for defining the standardized methods:

a^* = the alternative under analysis;

t = a unit of time, where $-t^a$ is the earliest point for alternative a (i.e., beginning of the study period) before the base year (i.e., $t = 0$), and T is the last point after the base year (i.e., end of the study period);

L = the length of the study period (e.g., $t^a + T$);

$B_t^{a^*}$ = the benefits for alternative a^* in year t;

$I_t^{a^*}$ = the investment costs for alternative a^* in year t;

$C_t^{a^*}$ = the non-investment costs for alternative a^* in year t;

$\underline{\mathbf{C}}_t^{a^*}$ = the combined cost for alternative a^* in year t (i.e., $\underline{\mathbf{C}}_t^{a^*} = I_t^{a^*} + C_t^{a^*}$);

$S_t^{a^*}$ = the savings for alternative a^* in year t; and

d = the discount rate expressed as a decimal.

Throughout this paper, the prefix, PV, is used to designate dollar denominated quantities in present value terms. The present value is derived by discounting (i.e., using the discount rate) to adjust all benefits, costs, and savings—past, present, and future—to the base year (i.e., $t = 0$). The dollar denominated quantities defined above and their associated present value terms are: the present value of benefits (PVB), the present value of investment costs (PVI), the present value of non-investment costs (PVC), the present value of combined costs (PV$\underline{\mathbf{C}}$), and the present value of savings (PVS).

PRESENT VALUE OF NET BENEFITS AND PRESENT VALUE NET SAVINGS

The present value of net benefits (PVNB) method is reliable, straightforward, and widely applicable for finding the economically efficient choice among alternatives (e.g., building systems). It measures the amount of net benefits from investing in a given alternative instead of investing in the foregone opportunity (e.g., some other alternative or maintenance of the status quo).

PVNB is computed by subtracting the time-adjusted costs of an investment from its time-adjusted benefits. If PVNB is positive, the investment is economic; if it is zero, the investment is as good as the next best investment opportunity; and if it is negative, the investment is uneconomical. Emphasis is on economic efficiency because the method is appropriate for evaluating alternatives that compete on benefits, such as revenue or other advantages that are measured in dollars, in addition to costs.

The present value of the net savings (PVNS) method is the PVNB method recast to fit the situation where there are no significant benefits in terms of revenue or the like, but there are reductions in future costs (e.g., reductions in the cost of ownership to consumers).[3] By treating savings like revenue benefits, the PVNB method may be reformulated as the PVNS method.

The PVNB for a given alternative, a^*, may be expressed as:

$$\begin{aligned} \text{PVNB}^{a^*} &= \text{PVB}^{a^*} - \text{PV}\underline{\text{C}^{a^*}} \\ &= \sum_{t=-t^a}^{T} \left(B_t^{a^*} - \underline{\text{C}_t^{a^*}}\right)/(1 + d^t) \end{aligned} \tag{1}$$

If there are no important benefits in terms of revenue, but there are reductions in future costs, then, the PVNS for a given alternative, a^*, may be expressed as:

$$\begin{aligned} \text{PVNS}^{a^*} &= \left(\text{PVS}^{a^*}\right) - \text{PVI}^{a^*} \\ &= \sum_{t=-t^a}^{T} \left(S_t^{a^*} - I_t^{a^*}\right)/(1 + d)^t \end{aligned} \tag{2}$$

If the decision maker anticipates revenues from the investment, then PVNB measure is used. If the decision maker expects costs to be reduced, then the PVNS measure is used. The PVNS measure is one of the methods used in the EPDM case study and is documented in ASTM E1074-09 [5].

BENEFIT-TO-COST RATIO AND SAVINGS-TO-INVESTMENT RATIO

The benefit-to-cost ratio (BCR) and the savings-to-investment ratio (SIR) are numerical ratios whose sizes indicate the economic performance of an investment. The BCR is computed as benefits, net of future non-investment costs divided by investment costs. The SIR is savings divided by investment costs. The SIR is the BCR method recast to fit the situation where the investment's primary advantage is lower costs. SIR is to BCR as PVNS is to PVNB.

A ratio less than 1.0 indicates an uneconomic investment; a ratio of 1.0 indicates an investment whose benefits or savings just equal its costs; and a ratio greater than 1.0 indicates an economic project. A ratio of, say, 4.75 means that the investor (e.g., the general public for a public-sector research program) can expect to receive $4.75 for every $1.00 invested (e.g., public funds expended), over and above the required rate of return imposed by the discount rate.

---

[3] If there are any benefits, say in the form of revenues or other positive cash flows, add them to the cost savings associated with the alternative under analysis.

The BCR for a given alternative, a^*, may be expressed as:

$$\mathrm{BCR}^{a^*} = \left(\mathrm{PVB}^{a^*} - \mathrm{PVC}^{a^*}\right)/\mathrm{PVI}^{a^*} = \frac{\sum_{t=-t^a}^{T} \left(B_t^{a^*} - C_t^{a^*}\right)/(1+d)^t}{\sum_{t=-t^a}^{T} I_t^{a^*}/(1+d)^t} \quad (3)$$

The SIR for alternative a^* may be expressed as:

$$\mathrm{SIR}^{a^*} = \mathrm{PVS}^{a^*}/\mathrm{PVI}^{a^*} = \frac{\sum_{t=-t^a}^{T} S_t^{a^*}/(1+d)^t}{\sum_{t=-t^a}^{T} I_t^{a^*}/(1+d)^t} \quad (4)$$

As was the case for the PVNB and PVNS measures, use the BCR if the decision maker anticipates revenues from the investment, and use the SIR if the decision maker anticipates costs to be reduced. The SIR measure is the second method used in the EPDM case study and is documented in ASTM E964-06 [6].

ADJUSTED INTERNAL RATE OF RETURN

The adjusted internal rate of return (AIRR) is the annual yield from a project over the study period, taking into account reinvestment of interim receipts. Because the AIRR calculation explicitly includes the reinvestment of all net cash flows, it is instructive to introduce a new term, terminal value (TV). The terminal value of an investment, a^*, is the future value (i.e., the value at the end of the study period) of reinvested net cash flows excluding all investment costs. The terminal value for an investment a^*, is denoted as TV^{a^*}.

The reinvestment rate in the AIRR calculation is equal to the minimum attractive rate of return (MARR), which is assumed to equal the discount rate, d, a constant. When the reinvestment rate is made explicit, all investment costs are easily expressible as a time equivalent initial outlay (i.e., a value at the beginning of the study period) and all non-investment cash flows (e.g., benefits, non-investment costs, savings) as a time equivalent terminal amount. This allows a straightforward comparison of the amount of money that comes out of the investment (i.e., the terminal value) with the amount of money put into the investment (i.e., the time equivalent initial outlay).

The AIRR is defined as the interest rate, r^*, applied to the terminal value, TV^{a^*}, which equates (i.e., discounts) it to the time equivalent value of the initial outlay of investment costs. It is important to note that all investment costs are discounted to a time equivalent initial outlay (i.e., to the beginning of the study period) using the discount rate, d.

Several procedures exist for calculating the AIRR. These procedures are derived and described in detail in the report by Chapman and Fuller [11]. The most convenient procedure for calculating the AIRR is based on its relationship to the BCR (SIR). This procedure results in a closed-form solution for r^*. The AIRR—expressed as a decimal—is that value of r^* for which:

$$\begin{aligned} r^* &= (1+d)(\mathrm{BCR}^{a^*})^{\frac{1}{L}} - 1 \\ &= (1+d)(\mathrm{SIR}^{a^*})^{\frac{1}{L}} - 1 \end{aligned} \tag{5}$$

The AIRR measure is the third method used in the SLP case study and is documented in ASTM E1057-06 [7].

SAVINGS IN WARRANTY REPAIR COSTS

The baseline analysis of the savings in warranty repair costs for high-performance seams for EPDM roofing is estimated as:

$$\begin{aligned} Z &= \sum_{t=t_0+1}^{T} \sum_{i=t-t_0}^{t_0} A_i * B_i * \frac{C_i}{D_i} * E_i * F * (1+R)^{T-i} \\ E_i &= \frac{B_i - B_{t_0}}{B_i} \end{aligned} \tag{6}$$

and

$$F = \frac{R(1+R)^{10}}{(1+R)^{10} - 1}$$

where:

Z = the present value of warranty cost savings in 2008 dollars,

A = the installed costs of EPDM roofing in millions of dollars (see **Table 1**),

B = the 10-year warranty cost as a percent of installed cost (see **Fig. 3**),

C = the 10-year warranty service cost for seams (see **Fig. 2**),

D = the 10-year indexed cost per m^2 (ft^2) (see **Fig. 1**),

E = the estimated proportion of 10-year warranty service costs for seams that can be averted because of improvements in technology,

F = the uniform capital recovery factor that converts estimated 10-year savings in warranty repair costs into a uniform annual stream of savings over a 10-year period,

R = the discount rate, and

t and i = index time.

Because this paper examines the 10-year warranty cost and has a 15-year study period; in this analysis, $t_0 = 9$ and $T = 24$.

To illustrate the conversion process, consider the year 1985. **Table 1** reports that the installed cost of EPDM roofs in 1985 was $\$1.55 \times 10^9$. **Figure 3** reports the 10-year warranty service costs as a percentage of original installed cost in 1985 as

6.7 %. Thus, the estimated 10-year warranty service costs for 1985 is $\$104.2 \times 10^6$. The seam warranty portion is 48.2 % [seam indexed cost per m^2 (ft^2) for 1985 of 11.8 (1.1) (see **Fig. 2**) divided by the indexed cost per m^2 (ft^2) for 1985 of 24.5 (2.3) (see **Fig. 1**)]. Thus, the estimated 10-year seam-related warranty service costs for 1985 is $\$50.3 \times 10^6$. The estimated proportion of 10-year warranty service costs for seams that can be averted because of improvements in technology is the difference between the value for B_i, in the case of 1985 6.7 %, and the value B_i for 1993, which is 1.1 %, divided by B_i, in this case 6.7 %. Thus, the estimated 10-year seam-related warranty service cost savings amounts to $42.0 million. Multiplying this amount by the uniform capital recovery factor for a 10-year period with a 7 % discount rate produces an annual seam-related warranty service cost savings estimate of $\$5.98 \times 10^6$. The $\$18.46 \times 10^6$ figure in column 2 of **Table 3** is the sum of the annual seam-related warranty service cost savings estimates for the years 1985 through 1993. **Table 3** summarizes the various stages in the calculations represented by Eq 6, including the year-by-year values.

TABLE 3 Summary of EPDM seam-related warranty repair savings: 1994 to 2008.

Year	Annual Dollar Amount of Warranty Cost Savings (in Millions of Current Dollars)	Conversion Factor by Year (Current Dollars to 2008 Dollars)	Annual Warranty Cost Savings by Year (in Millions of 2008 Dollars)	Single Present Value Factor by Year	Present Value of Annual Warranty Cost Savings by Year (in Millions of 2008 Dollars)	NIST Baseline Weighting Factor	Present Value of Annual Warranty Cost Savings Attributable to NIST by Year (in Millions of 2008 Dollars)
Column 1	Column 2	Column 3	Column 4 (2 × 3)	Column 5	Column 6 (4 × 5)	Column 7	Column 8 (6 × 7)
1994	18.46	1.453	26.82	2.579	69.16	0.25	17.29
1995	12.48	1.413	17.63	2.410	42.49	0.25	10.62
1996	9.03	1.372	12.39	2.252	27.92	0.25	6.98
1997	6.29	1.341	8.43	2.105	17.75	0.25	4.44
1998	4.39	1.321	5.80	1.967	11.41	0.25	2.85
1999	2.81	1.292	3.63	1.838	6.67	0.25	1.67
2000	1.25	1.250	1.57	1.718	2.69	0.25	0.67
2001	0.24	1.216	0.30	1.606	0.48	0.25	0.12
2002	0.00	1.197	0.00	1.501	0.00	0.25	0.00
2003	0.00	1.170	0.00	1.403	0.00	0.25	0.00
2004	0.00	1.140	0.00	1.311	0.00	0.25	0.00
2005	0.00	1.102	0.00	1.225	0.00	0.25	0.00
2006	0.00	1.068	0.00	1.145	0.00	0.25	0.00
2007	0.00	1.038	0.00	1.070	0.00	0.25	0.00
2008	0.00	1.000	0.00	1.000	0.00	0.25	0.00
Total					178.58	—	44.64

NIST's dual role as a facilitator via industry consortia and a research institution hastens the introduction of high-performance polymeric construction materials and expands their base of potential users. Although NIST is a key player, it is anticipated that the bulk of the costs savings are attributable to materials manufacturers and professional services within the construction industry. Thus, this paper, in keeping with its conservative approach, suggests a 25/75 split between NIST's contribution and that of other construction industry stakeholders. Such an accounting framework may be handled through use of a 0.25 weighting factor for NIST's contribution.

DISCOUNT RATE

The baseline analysis for the EPDM SLP economic impact assessment uses a real rate of 7 % to convert dollar amounts to present values. This rate is specified in Section 8.b of OMB Circular A-94 as the rate for all benefit-cost analyses of public investments and regulatory programs that provide benefits or incur costs to the general public [18].

Results

Three methods have been selected for examining NIST's investment in EPDM research: present value net savings, savings-to-investment ratio, and adjusted internal rate of return. These methods differ in the way in which they are calculated and, to some extent, in their applicability to particular types of investment decisions. The methods presented in the previous sections provide the basis for evaluating the economic performance of research investments. The equations underlying the methods presented earlier are all based on ASTM standard practices. All of the methods are appropriate for evaluating accept- or reject-type decisions.

The PVNS caused by improvements in the service life prediction of EPDM roofing that is attributable to NIST, expressed in millions of present value 2008 dollars and based on the approach outlined in the methods section, is equal to:

$$\begin{aligned} \text{PVNS} &= \text{PVS} - \text{PVcosts} \\ &= \$44.6 - \$17.9 \\ &= \$26.7\text{million} \end{aligned}$$

Utilizing the approach outlined previously, the SIR on NIST's EPDM-related investments is equal to:

$$\begin{aligned} \text{SIR} &= \text{PVS}/(\text{PVcosts}) \\ &= \$44.6/\$17.9 \\ &= 2.5 \end{aligned}$$

Utilizing the approach discussed earlier, the AIRR on NIST's EPDM-related investments is equal to:

$$\begin{aligned} \text{AIRR} &= (1 + 0.07) * 2.5^{1/15} - 1 \\ &= 0.137 \\ &= 13.7\% \end{aligned}$$

These results show that NIST's investment in EPDM research provided a positive net savings, making it an economical investment. It also has a savings-to-investment ratio greater than 1 and an adjusted internal rate of return higher than the minimum attractive rate of return, which is assumed to equal the discount rate.

Sensitivity Analysis

TYPES OF SENSITIVITY ANALYSIS AND WHY LATIN HYPERCUBE SAMPLING WAS SELECTED

Because the values of many variables that enter into the EPDM economic impact assessment are not known with certainty, it is advisable to select a small set of variables whose impact is likely to be substantial and subject them to a sensitivity analysis. Variations in the values of these input variables translate into the value of each outcome (e.g., the SIR) in such a manner that the impacts of uncertainty can be measured quantitatively. The sensitivity analysis employed in this paper is based on techniques for treating uncertainty and risk described in ASTM E1369-11 [8].

Sensitivity analysis may be divided into two polar cases: (1) deterministic, and (2) probabilistic. Deterministic sensitivity analyses are the most straightforward. Their advantage is that they are easy to apply and the results are easy to explain and understand. Their disadvantage is that they do not produce results that can be tied to probabilistic levels of significance (i.e., the probability that the SIR is less than 1.0).

For example, a deterministic sensitivity analysis might use as inputs a pessimistic value, a value based on a measure of central tendency (e.g., mean or median), and an optimistic value for the variable of interest. Then an analysis could be performed to see how each outcome (e.g., the SIR) changes as each of the three chosen values for the selected input is considered in turn, while all other input variables are maintained at their baseline values. A deterministic sensitivity analysis can also be performed on different combinations of input variables. That is, several variables are altered at once and then an outcome measure is computed. This is the approach used in two of the previous economic impact assessments [11].

In a probabilistic sensitivity analysis, a small set of key input variables is varied either singly or in combination according to an experimental design. In most cases, probabilistic sensitivity analyses are based on Monte Carlo or Latin hypercube techniques. The major advantage of probabilistic sensitivity analysis is that it permits the effects of uncertainty to be rigorously analyzed. For example, not only the expected value of each economic measure can be computed but also the variability of that value. In addition, probabilistic levels of significance can be attached to the computed values of each economic measure. The disadvantage of a probabilistic sensitivity analysis is that it requires many calculations carried out according to an experimental design, and is therefore practical only when used with a computer.

The approach selected for this study makes use of works by McKay et al. [19] and Harris [20]; it is based on the method of model sampling. Model sampling provides the basis for many probabilistic sensitivity analyses. Model sampling is a procedure for sampling from a stochastic process to determine, through multiple trials, the characteristics of a probability distribution. This approach was used in two of the previous economic impact assessments [21,22].

The method of model sampling was implemented through application of the Crystal Ball software [23]. This software product is an add-in for spreadsheets. For the case at hand, selected columns of the spreadsheet were associated with one or more of the nine input variables. Crystal Ball allows the user to specify a unique probability distribution for each input variable. Specification of the experimental design involves defining which variables are to be simulated and the number of simulations. Throughout this sensitivity analysis, 10 000 simulations were run for each combination of input variables under analysis. When the Crystal Ball software product is executed, it randomly samples from the parent probability distribution for each input variable of interest (i.e., the input variable(s) specified by the experimental design). In this analysis, a Latin hypercube sampling approach is used to ensure at least some sampling from the tails of the distributions. Latin hypercube sampling trades the randomness created by Monte Carlo sampling for sampling evenness. This yields better predictions (greater representativeness).

In reality, the exact nature of the parent probability distribution for each input variable is unknown. Estimates of the parameters (e.g., mean and variance) of the parent probability distribution can be made and uncertainty can be reduced by investigation and research. However, uncertainty can never be eliminated completely. The true specification of the parent probability distribution can only be known after EPDM products and services have been operating in the marketplace for an extended period of time. Therefore, to implement the procedure without undue attention to the characterization of the parent probability distribution, it was decided to focus on only two probability distributions. These probability distributions are: (1) the triangular, and (2) the uniform. Readers interested in learning more about these probability distributions, including variate relationships, estimation procedures, and random number generation, are referred to Evans et al. [24].

One reason for using these two probability distributions is that they are all defined over a finite interval. Furthermore, the specification of each probability distribution is accomplished with as few as two data points. The triangular distribution is widely used in simulation modeling; its specification requires three data points, the minimum value, the most likely value, and the maximum value. The triangular distribution is used whenever the range of input values is continuous and a clustering about some central value is expected. The uniform distribution is also widely used in simulation modeling; its specification requires only two data points, the minimum value and the maximum value. In addition, all values between the

TABLE 4 Assumptions for the EPDM Latin hypercube simulation.

Variable Name	Probability Distribution	Setting and Value		
		Most Likely	Minimum	Maximum
Discount rate	Uniform	0.07	0.04	0.10
Multiplier	Triangular	1	0.5	1.5

minimum and maximum are equally likely. The uniform distribution is used whenever the range of input values is continuous, but no a priori reason can be given for expecting clustering about some central value.

WARRANTY REPAIR COST SAVINGS AND ECONOMIC IMPACT

The present value of benefits of improved service life prediction (SLP) for the EPDM market is estimated to be $\$178.6 \times 10^6$, as seen in **Table 3**. The benefits attributed to NIST research are estimated to be $\$44.6 \times 10^6$. The baseline analysis assumed a 7 % discount rate. There is uncertainty surrounding the discount rate and the exact size of the benefits. In this uncertainty analysis, the discount rate follows a uniform distribution and varies between 4 % and 10 %, and the "true" (nominal) benefit is assumed to be 0.5 to 1.5 times the estimated ("multiplier") (see **Table 4**). Thus, given the degree of uncertainty, the actual warranty cost savings resulting from improved SLP in the EPDM market is between $\$69.3 \times 10^6$ and $\$368.3 \times 10^6$, as seen in **Fig. 4** and **Table 5**. The savings attributable to NIST research is somewhere between $\$17.3 \times 10^6$ and $\$92.1 \times 10^6$, as seen in **Table 5**. The PVNS is

FIG. 4 The simulated cumulative probability and frequency of the total savings because of improvements in EPDM.

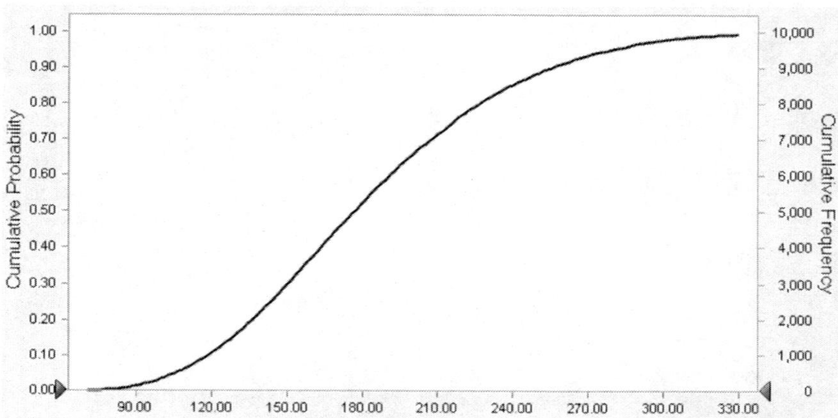

TABLE 5 Summary statistics of the Latin hypercube simulation because of NIST investments in EPDM.

	Total	Attributable to NIST			
Summary Statistics	Savings (in Millions of 2008 Dollars)	Savings (in Millions of 2008 Dollars)	PVNS (in Millions of 2008 Dollars)	SIR	AIRR
Trials	10 000	10 000	10 000	10 000	10 000
Mean	183.33	45.83	27.94	2.56	13.63
Median	177.60	44.40	26.51	2.48	13.68
Minimum	69.27	17.32	-0.58	0.97	6.77
Maximum	368.28	92.07	74.18	5.15	19.35
Standard deviation	52.64	13.16	13.16	0.74	2.20
Mean standard error	0.53	0.13	0.13	0.01	0.02

between −0.6 and 74.2, the SIR is between 1.0 and 5.2, and the AIRR is between 6.8 and 19.4. Although the worst-case scenario has a negative PVNS and an SIR less than 1, the cumulative probability of a negative PVNS and an SIR less than 1 is 0.0003.

Using Latin hypercube sampling, the distribution of the two uncertainty parameters were sampled 10 000 times (trials). This produced 10 000 benefit estimates. The cumulative distribution for the results are graphed in **Figs. 4** through **8** with **Table 5** being the summary information for the data. The lowest value of

FIG. 5 The simulated cumulative probability and frequency of the warranty savings because of NIST investments in EPDM.

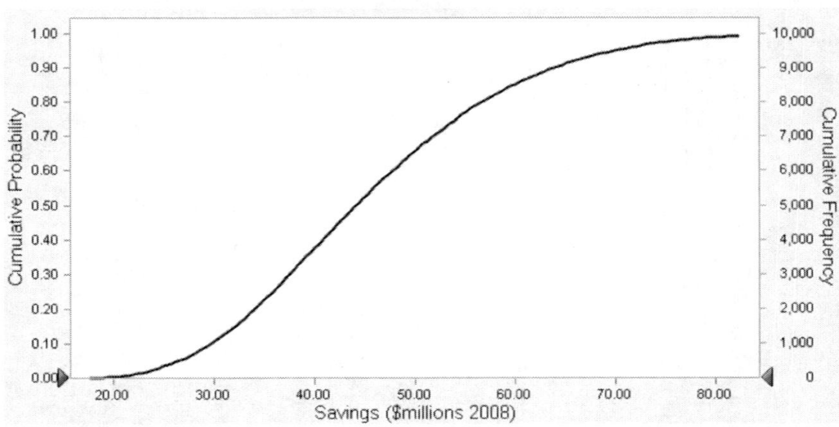

FIG. 6 The simulated cumulative probability and frequency of the PVNS because of NIST investments in EPDM.

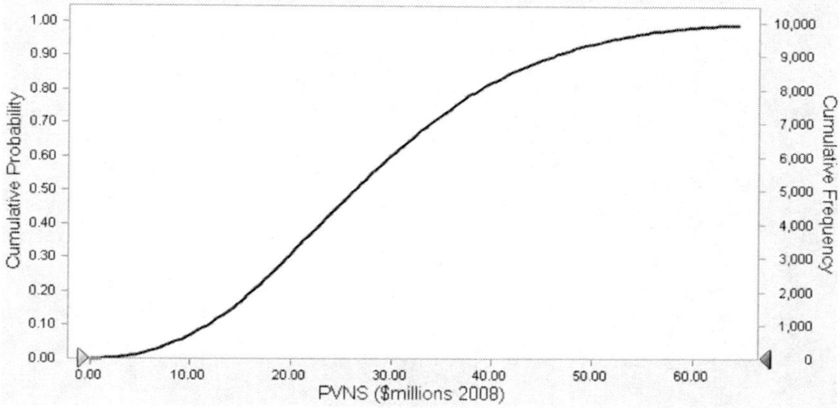

savings attributable to NIST ($\$17.3 \times 10^6$) is at the left far end of the cumulative graph in **Fig. 5**. As one moves from the left to right along the graph, it gives the probability that the savings is below the corresponding value and shows the proportion (frequency) of simulated trials that produce a savings below each value. The point on the graph where the cumulative probability equals 0.5, which yields a savings of $\$44.4 \times 10^6$, represents the point where there is a 0.5 probability of the savings attributable to NIST being $\$44.4 \times 10^6$ or less. The probability that the PVNS,

FIG. 7 The simulated cumulative probability and frequency of the SIR because of NIST investments in EPDM.

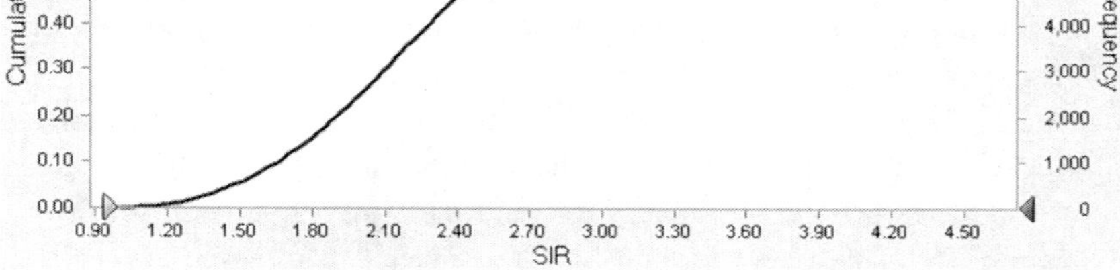

FIG. 8 The simulated cumulative probability and frequency of the AIRR because of NIST investments in EPDM.

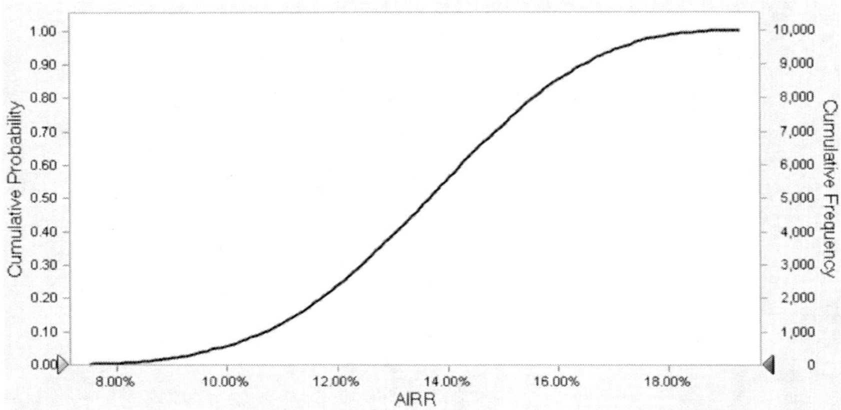

SIR, and AIRR are 26.7×10^6, 2.5, and 13.7 %, respectively, the previously estimated values, is 0.51.

Summary

This case study estimates the economic impacts to the non-residential buildings sector of the construction industry from NIST's EPDM-related research. This economic impact assessment covers the period from 1994 through 2008. Therefore, it is an ex post or retrospective economic impact assessment. The reduction in warranty service repair costs is a substantial benefit from better-performing EPDM roofing systems and is the focus of this paper. The total present value of warranty cost savings is estimated to be 178.6×10^6 (2008 dollars) with 44.6×10^6 being attributed to NIST. With a present value investment of 17.9×10^6, the resulting PVNS is 26.7×10^6 with an SIR of 2.5 and AIRR of 13.7 %, suggesting that the NIST research investment in EPDM roofing resulted in an economical investment.

Because the values of many variables that enter into the EPDM economic impact assessment are not known with certainty, a small set of variables were used in a sensitivity analysis. Variations in the values of these input variables translate into the value of each outcome (e.g., the SIR) in such a manner that the impacts of uncertainty can be measured quantitatively. The baseline analysis assumed a 7 % discount rate. There is uncertainty surrounding the discount rate and the exact size of the benefits. In this uncertainty analysis the discount rate follows a uniform distribution and varies between 4 % and 10 %, and the "true" (nominal) benefit is assumed to be 0.5 to 1.5 times the estimated ("multiplier"). Thus, given the degree of uncertainty, the actual warranty cost savings resulting from improved SLP in the

EPDM market is between \$69.3 × 10^6 and \$368.3 × 10^6. The savings attributable to NIST research is somewhere between \$17.3 × 10^6 and \$92.1 × 10^6. The PVNS is between -\$0.6 × 10^6 and \$74.2 × 10^6, the SIR is between 1.0 and 5.2, and the AIRR is between 6.8 % and 19.4 %. Although the worst-case scenario has a negative PVNS and an SIR less than 1, the cumulative probability of a negative PVNS and an SIR less than one is 0.0003.

The "case study" approach employed here illustrates how to evaluate and compare the economic impacts of research investments. Standardized methods are used in this paper to ensure consistency in the measurement of economic impacts. The benefits and cost savings calculated in this paper are a lower-bound estimate, because our conservative approach included only those that are clearly definable and measurable.

ACKNOWLEDGMENTS

The policy of the National Institute of Standards and Technology and of ASTM International is to use metric units in all of their published materials. Because this paper is intended for the U.S. construction industry that uses U.S. customary units, it is more practical and less confusing to include U.S. customary units as well as metric units. Measurement values in this paper are therefore stated in metric units first, followed by the corresponding values in U.S. customary units in parentheses. Certain trade names and company products are mentioned in the text to adequately specify the technical procedures and equipment used. In no case does such identification imply recommendation or endorsement by the National Institute of Standards and Technology, nor does it imply that the products are necessarily the best available for the purpose.

References

[1] Griffin, C. W., *Manual of Built-up Roof Systems*, McGraw-Hill, New York, 1970, p. 1.

[2] Mathey, R. G. and Cullen, W. C., "Preliminary Performance Criteria for Bituminous Membrane Roofing," *Building Science Series 55*, National Bureau of Standards, Washington, D.C., 1974.

[3] Chapman, R. E., Butry, D. T., Huang, A. L., and Thomas, D. S., "Benefits and Costs of Research: A Case Study of Improved Service Life Prediction," *NIST Technical Note 1650*, National Institute of Standards and Technology, Gaithersburg, MD, 2009.

[4] National Roofing Contractors Association, "Annual Market Survey: 2006–2007," National Roofing Contractors Association, Rosemont, IL, 2007.

[5] ASTM E1074-09: Practice for Measuring Net Benefits and Net Savings for Investments in Buildings and Building Systems, *Annual Book of ASTM Standards*, ASTM International, West Conshohocken, PA, 2012.

[6] ASTM E964-06: Practice for Measuring Benefit-to-Cost and Savings-to-Investment Ratios for Buildings and Building Systems, *Annual Book of ASTM Standards*, ASTM International, West Conshohocken, PA, 2012.

[7] ASTM E1057-06: Standard Practice for Measuring Internal Rate of Return and Adjusted Internal Rate of Return for Investments in Buildings and Building Systems, *Annual Book of ASTM Standards*, ASTM International, West Conshohocken, PA, 2012.

[8] ASTM E1369-11: Standard Guide for Selecting Techniques for Treating Uncertainty and Risk in the Economic Evaluation of Buildings and Building Systems, *Annual Book of ASTM Standards*, ASTM International, West Conshohocken, PA, 2012.

[9] Hoff, J. L., 2003, "An Update of Historical Warranty Service Costs," RCI Interface, http://www.rci-online.org/interface/2003-09-hoff.pdf (Last accessed 1 Sept 2009).

[10] ASTM, *ASTM Standards on Building Economics*, 7th ed., ASTM International, West Conshohocken, PA, 2012.

[11] Chapman, R. E. and Fuller, S. K., 1996, "Two Case Studies in Building Technology," NISTIR 5840, pp. 27–37, http://fire.nist.gov/bfrlpubs/build96/PDF/b96113.pdf (Last accessed 1 March 2014).

[12] Ruegg, R. T. and Marshall, H. E., *Building Economics: Theory and Practice*, Chapman & Hall, London, 1990.

[13] U.S. Census Bureau, "1982 Census of Construction Industries," Industry Series, Suitland, MD, 1984.

[14] U.S. Census Bureau, "1987 Census of Construction Industries," Industry Series, Suitland, MD, 1990.

[15] U.S. Census Bureau, "1992 Census of Construction Industries," Industry Series, Suitland, MD, 1996.

[16] U.S. Census Bureau, "1997 Census of Construction Industries," Industry Series, Suitland, MD, 2000.

[17] National Roofing Contractor Association, "Project Pinpoint Analysis: Ten-Year Performance Experience of Commercial Roofing 1983–1992," National Roofing Contractors Association, Rosemont, IL, 1993.

[18] Executive Office of the President, OMB Circular A-94, Washington, D.C., 1992.

[19] McKay, M. C., Conover, W. H., and Beckman, R. J., "A Comparison of Three Methods for Selecting Values of Input Variables in the Analysis of Output from a Computer Code," *Technometrics*, Vol. 21, 1979, pp. 239–245.

[20] Harris, C. M., "Issues in Sensitivity and Statistical Analysis of Large-Scale, Computer-Based Models," *NBS GCR 84-466*, National Bureau of Standards, Gaithersburg, MD, 1984.

[21] Chapman, R. E., 2000, "Benefits and Costs of Research: A Case Study of Construction Systems Integration and Automation Technologies in Industrial Facilities," NISTIR 6501, http://fire.nist.gov/bfrlpubs/build00/PDF/b00025.pdf (Last accessed 1 March 2014).

[22] Chapman, R. E., 1999, "Benefits and Costs of Research: A Case Study of Cybernetic Building Systems," NISTIR 6303, http://fire.nist.gov/bfrlpubs/build99/PDF/b99003.pdf (Last accessed 1 March 2014).

[23] *Crystal Ball 11.1.2.3; User's Manual.* (2013). Decisioneering, Denver, CO.

[24] Evans, M., Hastings, N., and Peacock, B., *Statistical Distributions*, Wiley, New York, 1993.

STP 1586, 2014 / available online at www.astm.org / doi: 10.1520/STP158620140038

David T. Butry[1]

The Economics of Residential Fire Sprinklers and the Potential Impact of Recent Code Changes

Reference

Butry, David T., "The Economics of Residential Fire Sprinklers and the Potential Impact of Recent Code Changes," *Ideas to Impact: How Building Economic Standards Keep You on Track*, STP 1586, Robert E. Chapman and Michael N. Goodkind, Eds., pp. 149–163, doi:10.1520/STP158620140038, ASTM International, West Conshohocken, PA 2014.[2]

ABSTRACT

Residential fire sprinklers are expected to prevent room flashover, improving conditions for occupant egress and thereby reducing the number of fatalities resulting from home fires. Based on data reported by the National Fire Protection Association, in the U.S., occupant deaths resulting from fires in one- and two-family dwellings ("homes") averaged 2154 between 2008 and 2012—some 71 % of all fire deaths. Of these fires, none involved homes with sprinklers, partly due to their effectiveness, but also due to their rarity in the U.S. housing stock. Their scarcity is largely explained by economic considerations. Using ASTM Standard Practices, we demonstrate the economic performance of sprinklers is primarily driven by the (1) physical performance of sprinklers in terms of fatalities avoided, (2) value of a statistical life (VSL), (3) time value of money (discount rate), (4) insurance savings for owners of sprinklered homes, and (5) sprinkler installation costs. Only until recently have the benefits been shown to outweigh the costs, specifically when the VSL is high and the discount rate is low. Such research was cited in support of a change to the International Residential Code requiring sprinklers in all new one- and two-family homes, which became effective Jan. 1, 2011. We evaluate the impact—in terms of occupant fatalities and occupant and fire service injuries avoided—from continued penetration of sprinklers into the U.S. housing stock. The results

Manuscript received April 2, 2014; accepted for publication April 14, 2014; published online July 14, 2014.

[1]Economist, Applied Economics Office, Engineering Laboratory, National Institute of Standards and Technology, Gaithersburg, MD 20899.

[2]ASTM Symposium on *Ideas to Impact: How Building Economic Standards Keep You on Track* on October 5, 2014 in New Orleans, LA.

suggest that the impact can be significant, but is moderated by the rate of statewide code adoption and the turnover in the housing market.

Keywords
fire sprinklers, economics, market penetration, sprinkler requirements

Introduction

Residential fire sprinklers are expected to prevent room flashover, improving conditions for occupant egress and thereby reducing the number of fatalities resulting from home fires (see [1]). Based on data reported by the National Fire Protection Association, in the U.S., fires in one- and two-family homes ("homes") averaged 277 000 annually from 2008 to 2012, representing 20 % of all reported fires (see Refs. [2,3]). Yet occupant deaths resulting from home fires averaged 2154 annually, representing 71 % of all fire deaths (see Refs. [2,3]). Of these fires, none involved homes with fire sprinklers, partly due to their effectiveness, but also possibly due to their rarity in the U.S. housing stock. Butry et al. [4] show the number of fatalities and injuries are far fewer for home fires involving sprinklers than those without—a 100 % reduction in occupant fatalities and a 57 % reduction in occupant injuries has been observed. However, the number of home fires involving sprinklers is rare. Hall [5] reports that only 2.2 % of all occupied single-family homes in the U.S. had fire sprinklers in 2009, although the percentage is up from previous years and expected to continue to grow.

Table 1 shows the number of reported fires, occupant fatalities, and occupant injuries associated with home fires, and the number of on-scene fire service injuries, from 2008 to 2012. Given the life-safety performance of fire sprinklers and their relative scarcity in the housing market, increased sprinkler use has great potential to reduce a significant portion of fire deaths and injuries – sprinklers have been shown to operate during 91 % of all structure fires, when present, and of these, they are 96 % effective in containing and controlling the fire until the fire department can complete the extinguishment [4]. Their presence in the nationwide housing stock is moderated by two factors: the adoption of sprinkler requirements for new home construction by states and local jurisdictions, and the rate of turnover (replacement of pre-existing homes) in the housing market.

The objectives of this paper are (1) to demonstrate, using ASTM Standard Practices, those economic factors that affect the benefit-cost performance (present value net benefits) of residential fire sprinklers, from the homeowner's perspective, and (2) to evaluate the total impact, in terms of occupant fatalities and occupant and fire service injuries avoided, from continued penetration of sprinklers into the U.S. housing stock.

Methods

Two methods are described: (1) the present value net benefits method used to evaluate the benefit-cost performance of residential fire sprinklers, and (2) an impact

TABLE 1 Number of reported fires, occupant (O) fatalities, and occupant (O) injuries in one- and two-family homes [2], and number of on-scene fire service (FS) injuries of all incident types [6], 2008 to 2012.

Year	Fires	Fatalities (O)	Injuries (O)	Injuries (FS)
2008	291 000	2365	9185	36 595
2009	272 500	2100	9300	32 205
2010	279 000	2200	9400	32 675
2011	274 500	2105	9485	30 505
2012	268 000	2000	12 875	31 490
Average	277 000	2154	10 049	32 694

projection method used to estimate the number of fatalities and injuries avoided from increased market penetration of sprinklered homes.

PRESENT VALUE NET BENEFITS

The benefits experienced by occupants of sprinklered homes, as described in this paper, include reductions in the following: the risk of civilian fatalities and injuries, homeowner insurance premiums, uninsured direct property losses, and uninsured indirect costs. The primary costs examined are the initial purchase and installation of the sprinkler system. The measure of benefit-cost performance, the present value net benefits, compares differently timed benefit and cost cash flows, accruing to a homeowner, by discounting them to a reference point in time. All dollars presented in this paper are in 2013 adjusted dollars. The present value net benefits are calculated by subtracting present value costs from the present value benefits.

The quantified benefits of a fire sprinkler system used in a single-family dwelling are fully described in Butry et al. [4], based on reported fire incident data contained within the U.S. Fire Administration's National Fire Incident Reporting System scaled to national estimates using reported data based on the National Fire Protection Association's annual survey of fire departments over the period from 2002 to 2005. Over this period, homes equipped with smoke alarms and a fire sprinkler system experienced 100 % fewer occupant fatalities, 57 % fewer occupant injuries, and 32 % less direct property losses and indirect costs resulting from fire than homes equipped only with smoke alarms. In addition, it was common for homeowners of dwellings with fire sprinkler systems to receive an 8 % reduction in their homeowner insurance.

The costs of a fire sprinkler system are based on Brown [7], which documents the design and installation costs of four different fire sprinkler systems in three prototypical house types. The four sprinkler configurations included one multipurpose network system (tied into the domestic plumbing) and three stand alone designs, with and without a backflow preventer. The three house types are a Colonial (3338 ft^2; 310 m^2), Ranch (1171 ft^2; 109 m^2), and a Townhouse (2257 ft^2; 210 m^2). In this paper, all three house designs are considered, while only the multipurpose network system is used as it is shown to be the most cost-effective.

The generalized present value of net benefits equation, PVNB, adapted from ASTM E1074-09 [8], for comparing benefits and costs attributed to the addition of a sprinkler system to a house with only smoke alarms, is given by:

$$PVNB = \sum_{t=0}^{T} \frac{B_t - C_t}{(1+d)^t} \tag{1}$$

where:

B_t = the dollar value of benefits in period t,

C_t = the dollar value of costs in period t,

T = the number of discounting time periods in the study period, and

d = the discount rate per time period.

A positive PVNB implies the present value benefits outweigh the present value costs. The discount adjustment, $(1+d)^{-t}$, is consistent with discount formulas found in the adjunct to ASTM E917-05 [9].

In the case of sprinklers, the benefit and cost terms from Eq 1 can be decomposed into individual components.

Present value benefits, PVB, can be expressed as:

$$PVB = \sum_{t=0}^{T} \frac{B_t}{(1+d)^t} = \sum_{t=0}^{T} \frac{D_t + I_t + U_t + N_t}{(1+d)^t} \tag{2}$$

where:

D = the statistical value of death and injury averted due to sprinkler use,

I = the value of reduced uninsured direct and non-reimbursable property loss due to sprinkler use,

U = the value of reduced uninsured indirect and non-reimbursable costs due to sprinkler use, and

N = the value of reduced homeowner insurance premiums due to the presence of sprinklers.

The present value costs, PVC, can be expressed as:

$$PVC = C_0 + \sum_{t=1}^{T} \frac{C_t}{(1+d)^t} \tag{3}$$

where:,

C_0 = the purchase and installation cost of a sprinkler system, and

C_t = the operations and maintenance-related costs, which tend to be minimal.

Other benefits and costs might exist, but they are not considered here (note: dividing any one of the individual benefits or costs by $(1+d)^t$ yields its individual present value).

This paper reports the present value net benefits found in Butry et al. [4] and extends that research using break-even analysis adapted from ASTM E1369-11 [10].

The break-even analysis provides insight into the sensitivity of the original results by examining the changes needed (to the baseline assumptions) to equate the present value benefits to the present value costs.

IMPACT PROJECTIONS

Beginning with the 2009 International Residential Code (IRC) [11], approved sprinklers are required in all new one- and two-family dwellings (including townhouses) beginning in 2011[3]. Currently the code has been adopted statewide by California and Maryland.

Butry [12] measures differences in the rate of occupant fatalities, occupant injuries, and fire service injuries between sprinklered and non-sprinklered homes, but unlike Butry et al. [4] and Butry [13], held constant differences in neighborhood socio-economics (e.g., race, age, population size) and housing characteristics (e.g., age, vacancy rates, type of smoke alarm) to reduce potential confounding effects. The results confirmed the estimates used in the previous research. The findings show that sprinklers reduce occupant fatalities by 100 % (95 % confidence interval: ±29 %), occupant injuries by 45 % (95 % confidence interval: ±44 %), and fire service injuries by 87 % (95 % confidence interval: ±34 %).

The number of sprinklered homes is growing, and because the recent IRC sprinkler code requirement will enhance this growth, we examine the potential impact sprinklers could have, using the results from Butry [12], over a range of market penetration rates—i.e., percentage of housing stock with sprinklered homes.

Potential sprinkler impact is estimated as:

$$A_{k,t} = F_k * H_k * S_t \tag{4}$$

where:

A = the number of adverse health outcomes averted due to sprinkler use,

F = the % reduction in adverse health outcome due to sprinkler use,

H = the number of adverse health outcomes occurring with no sprinklered homes in the housing stock,

S = the % of the housing stock with sprinklers, and

k = indexes adverse health outcomes (occupant fatalities, occupant injuries, fire service injuries).

Results

PRESENT VALUE NET BENEFITS OF RESIDENTIAL FIRE SPRINKLERS

Table 2 shows the inputs used to measure the present value net benefits of residential fire sprinklers installed in new homes. For each of the prototypical homes

[3] In the IRC code proposal establishing the sprinkler requirement (RB64 07-08), Butry et al. [4] was cited as evidence of the potential fire sprinklers can have reducing fire fatalities.

TABLE 2 Inputs and parameters for the present value net benefits calculation of residential fire sprinklers.

Inputs	Label	Value	Present Value
Benefits			
Fatalities and Injuries Averted	D	$298.87	$4700.87
Direct Property Losses Averted	I	$6.02	$94.77
Indirect Costs Averted	U	$1.20	$18.96
Insurance Credit	N	$71.76	$1128.61
Costs			
Installation	P		
Colonel		$2469.35	$2469.35
Ranch		$986.11	$986.11
Townhouse		$2255.25	$2255.25
Parameters			
Study (Time) Period	T	30	N/A
Discount Rate	d	4.8 %	N/A

considered, residential fire sprinklers produce positive present value net benefits: $3474 for the Colonial, $4957 for the Ranch, and $3688 for the Townhouse.

It is clear from the **Table 2** that the economic performance of fire sprinklers is driven by the value of preventing occupant fatalities and injuries, insurance savings for owners of sprinklered homes, and sprinkler installation costs, as these are the three largest values. The economic value of life-safety is heavily influenced by the assumed value of a statistical life and injury (see Butry [4] for more details). Not apparent in **Table 2** is the sensitivity of the results to changes to the study period and discount rate. Because the economic benefits of sprinklers accrue over the lifetime of the equipment, while the purchase and installation costs occur at the time of home construction, (1) increasing the study period results in an increase to the benefits only, unless the study period is increased so long that the sprinkler system requires repairs or replacement, and (2) increasing (decreasing) the discount rate decreases (increases) the value of future benefits.

Table 3 presents the results of the break-even analysis used to show the required input values needed to cause the present value benefits to equal the present value costs. Baseline values (those that correspond with **Table 2**) are shown for comparison. For each of the home types, an installation cost of $5943 is needed to create break-even. For the Colonial (3338 ft^2; 310 m^2) this would require an increase in the installation cost from $0.74/ft^2 to $1.78/ft^2 ($7.97/m^2 to $19.16/m^2). Similarly, a $5943 installation cost would require a per unit price increase from $0.84/ft^2 to

TABLE 3 Input values required to achieve break-even (present value benefits equal to present value costs), all else equal.

Inputs	Baseline	Colonial	Ranch	Townhouse
Installation Cost ($)	Varies	5943	5943	5943
Value of a Statistical Life ($million)[a]	9.4	2.0	0[b]	1.5
Reduction in Fatalities (%)[a]	100	21.6	0[b]	16.8
Fires per 10 000 homes[a]	36	10	0[b]	9
Reduction in Insurance (%)	8	0[b]	0[b]	0[b]
Study Period (Years)	30	8.0	2.8	7.2
Discount Rate (%)	4.8	15.1	38.3	16.6

[a]Factors affecting the "Fatalities and Injuries Averted" input value.
[b]Break-even could not be achieved even with a value of 0. With a value of 0, present value benefits exceeded present value costs.

$5.08/ft^2 ($9.04/m^2 to $54.68/m^2) for the Ranch (1171 ft^2; 109 m^2) and an increase from $1.00/ft^2 to $2.63/ft^2 ($10.76/m^2 to $28.31/m^2) for the Townhouse (2257 ft^2; 210 m^2).

The value of a statistical life (VSL), reduction in fire-related fatalities due to sprinkler use, and the fire occurrence per 10 000 homes are used to value the benefits from reducing fatalities and injuries. The baseline VSL assumes a statistical life is worth $\$9.4 \times 10^6$. The break-even analysis reveals that a VSL of $\$2.0 \times 10^6$ for the Colonial and $\$1.5 \times 10^6$ for the Townhouse is needed to create break-even. Even with a VSL of $0, a break-even point could not be achieved for the Ranch. This implies residential fire sprinklers, in homes with a cost and insurance savings similar to the Ranch prototype, are economic even without considering their life-safety benefits (for the Ranch, the installation costs are completely offset by the present value insurance savings. This explains why a break-even point could not be achieved for VSL, % reduction in fatalities, and fire occurrence per 10 000 homes).

In the baseline analysis, fire sprinklers are assumed to reduce the number of fire-related deaths by 100 %, as no fire fatalities have been reported of occupants in sprinklered one- and two-family dwellings. However, **Table 3** shows that the life-safety performance could be as low to produce a 21.6 % reduction in the Colonial and a 16.8 % reduction in the Townhouse for the present value benefits to equal the present value costs, holding all other inputs at their baseline level. In addition, the likelihood of home fire could be reduced from 36 fires per 10 000 homes (annually) to 10 fires per 10 000 for the Colonial and 9 fires per 10 000 for the Townhouse for the benefits to completely offset the costs, meaning in areas of the country with home fire risks up to 72 % (Colonial) to 75 % (Townhouse) less than the national average, fire sprinklers still yield positive net economic benefits (assuming all other inputs are similar to the baseline assumed).

It appears so long as homeowners receive economic benefits from life-safety, similar to the baseline assumptions, the insurance savings could be $0/year and fire sprinklers would still be economic, as the difference between the present value benefits and costs exceed the baseline present value insurance savings. Varying the study period and applying ASTM E1121-12 [14] demonstrates that fire sprinklers pay for themselves, using the prototypes considered, within three to eight years, all else equal. Finally, different discount rates were considered, varying between 15.1 % for the Colonial and 38.3 % for the Ranch before present value benefits equaled present value costs.

Having shown the benefit-cost considerations driving the economic performance of fire sprinklers, we turn attention to the impact sprinklers can have with increased market penetration in the U.S. housing stock.

IMPACT OF INCREASED MARKET PENETRATION OF FIRE SPRINKLERS

Hall [4] reports that 2.2 % of all occupied single-family homes (attached and detached) had fire sprinklers in 2009, up from 1.9 % in 2007. Assuming a similar

TABLE 4 Annual reduction and percent reduction of occupant fatalities at different market penetration level (percent of housing stock) for single family housing, and their 95 % confidence intervals.

Market	Reduction			Percent Reduction		
Penetration (%)	Lower	Expected	Upper	Lower	Expected	Upper
1	15	22	22	1	1	1
2	31	43	43	1	2	2
3	46	65	65	2	3	3
4	62	86	86	3	4	4
5	77	108	108	4	5	5
6	92	129	129	4	6	6
7	108	151	151	5	7	7
8	123	172	172	6	8	8
9	139	194	194	6	9	9
10	154	215	215	7	10	10
11	169	237	237	8	11	11
12	185	258	258	9	12	12
13	200	280	280	9	13	13
14	216	302	302	10	14	14
15	231	323	323	11	15	15
16	246	345	345	11	16	16
17	262	366	366	12	17	17
18	277	388	388	13	18	18
19	293	409	409	14	19	19
20	308	431	431	14	20	20

increase in market penetration of 0.3 percentage points every two years, and ignoring any impact of recent IRC adoption in California and Maryland, this would imply 2.8 % of all occupied single family homes were sprinklered in 2013 (this serves as a lower bound estimate). Based on the American Housing Survey's, Components of Inventory Change study [14], new homes account for approximately 1 % of the housing stock each year. If the entire U.S. required sprinklers in all new single-family homes, beginning in 2014, in ten years, sprinklered homes would comprise 12.8 % of the housing stock. This, of course, represents an upper limit, as nationwide state adoption of the sprinkler requirement has not occurred (it also assumes no loss in the number of new homes built over this period, which, again, makes 12.8 % an upper limit). Therefore, as states and local jurisdiction mandate sprinklers over the next ten years, the level of market penetration is likely to be somewhere between 2.8 % (current) and 12.8 % (maximum).

Tables 4–6 provide the estimated annual impacts on occupant fatalities (**Table 4**), occupant injuries (**Table 5**), and fire service injuries (**Table 6**) for different market

TABLE 5 Annual reduction and percent reduction of occupant injuries at different market penetration level (percent of housing stock) for single family housing, and their 95 % confidence intervals.

Market	Reduction			Percent Reduction		
Penetration (%)	Lower	Expected	Upper	Lower	Expected	Upper
1	1	45	90	0	0	1
2	3	91	179	0	1	2
3	4	136	269	0	1	3
4	6	182	358	0	2	4
5	7	227	448	0	2	4
6	8	273	537	0	3	5
7	10	318	627	0	3	6
8	11	364	716	0	4	7
9	13	409	806	0	4	8
10	14	455	895	0	5	9
11	16	500	985	0	5	10
12	17	546	1075	0	5	11
13	18	591	1164	0	6	12
14	20	637	1254	0	6	12
15	21	682	1343	0	7	13
16	23	728	1433	0	7	14
17	24	773	1522	0	8	15
18	25	819	1612	0	8	16
19	27	864	1701	0	9	17
20	28	910	1791	0	9	18

penetration levels. Each table shows the expected reduction in fatalities and injuries and % reduction in fatalities and injuries, along with their associated upper and lower bounds based on computed 95 % confidence intervals (95 % CI). Note that the upper bound for fatalities averted equals the expected value because fatalities cannot be reduced beyond 100 %.

Currently (3 % market penetration), sprinklers are estimated to reduce 65 (95 % CI: 46 to 65) occupant fatalities, 136 occupant injuries (95 % CI: 4 to 269), and 853 fire service injuries (95 % CI: 523 to 981) each year. These reductions constitute a 3 % reduction (95 % CI: 2 to 3 %) in occupant fatalities, a 1 % reduction (95 % CI: 0 to 3 %) in occupant injuries, and a 3 % reduction (95 % CI: 2 to 3 %) in fire service injuries.

At 10 % market penetration, sprinklers are expected to reduce *an additional* 150 (95 % CI: 108 to 150) occupant fatalities, 319 occupant injuries (95 % CI: 10 to 626), and 1991 fire service injuries (95 % CI: 1219 to 2288) each year. These reductions constitute *an additional* 7 percentage point reduction (95 % CI: 5 to 7) in

TABLE 6 Annual reduction and percent reduction of fire service injuries at different market penetration level (percent of housing stock) for single-family housing, and their 95 % confidence intervals.

Market	Reduction			Percent Reduction		
Penetration (%)	Lower	Expected	Upper	Lower	Expected	Upper
1	174	284	327	1	1	1
2	348	569	654	1	2	2
3	523	853	981	2	3	3
4	697	1137	1308	2	3	4
5	871	1422	1635	3	4	5
6	1045	1706	1962	3	5	6
7	1219	1991	2289	4	6	7
8	1394	2275	2616	4	7	8
9	1568	2559	2942	5	8	9
10	1742	2844	3269	5	9	10
11	1916	3128	3596	6	10	11
12	2090	3412	3923	6	10	12
13	2265	3697	4250	7	11	13
14	2439	3981	4577	7	12	14
15	2613	4265	4904	8	13	15
16	2787	4550	5231	9	14	16
17	2962	4834	5558	9	15	17
18	3136	5118	5885	10	16	18
19	3310	5403	6212	10	17	19
20	3484	5687	6539	11	17	20

FIG. 1 Projected number of occupant fatalities averted accruing over time, with associated 95 % confidence interval, from residential fire sprinklers, using different annual market penetration rates (AMPR), starting with 3 % of all single family homes sprinklered in Year 1.

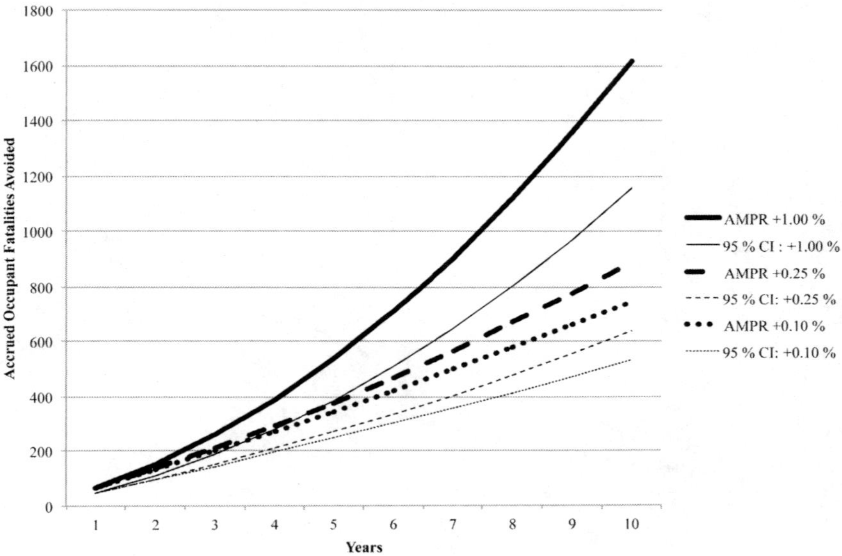

occupant fatalities, 4 percentage point reduction (95 % CI: 0 to 6) in occupant injuries, and a 6 percentage point reduction (95 % CI: 3 to 7) in fire service injuries.

Figures 1–3 show the projected number of occupant fatalities (**Fig. 1**), occupant injuries (**Fig. 2**), or fire service injuries (**Fig. 3**) accruing over a ten-year period from continued market penetration of fire sprinklers in one- and two-family homes. Year 1 begins with 3 % of all homes being sprinklered, and three different annual market penetration rate increases (AMPR) are considered: 0.10, 0.25, and 1.00 annual percentage point increases. The 95 % confidence interval is provided for each projection (due to truncation, the upper confidence interval corresponds with the expected value for occupant fatalities, as greater than 100 % reduction in fatalities, per fire, cannot be achieved).

In the figures, growth in the number of sprinklered homes can come from (1) growth in the number of sprinklered jurisdictions, (2) continued home turnover in sprinkler jurisdictions, or (3) both. The figures are non-specific in the nature of the penetration growth. However, with a few assumptions, regarding its nature, the figures provide additional insight. If the turnover rate and the number of sprinklered jurisdictions are assumed fixed over time, with the turnover rate set to 1 % (the current estimate), then the figures can be used to show the expected number of outcomes (fatalities or injuries avoided) for when the % of U.S. housing stock with a

FIG. 2 Projected number of occupant injuries averted accruing over time, with associated 95 % confidence interval, from residential fire sprinklers, using different annual market penetration rates (AMPR), starting with 3 % of all single family homes sprinklered in Year 1.

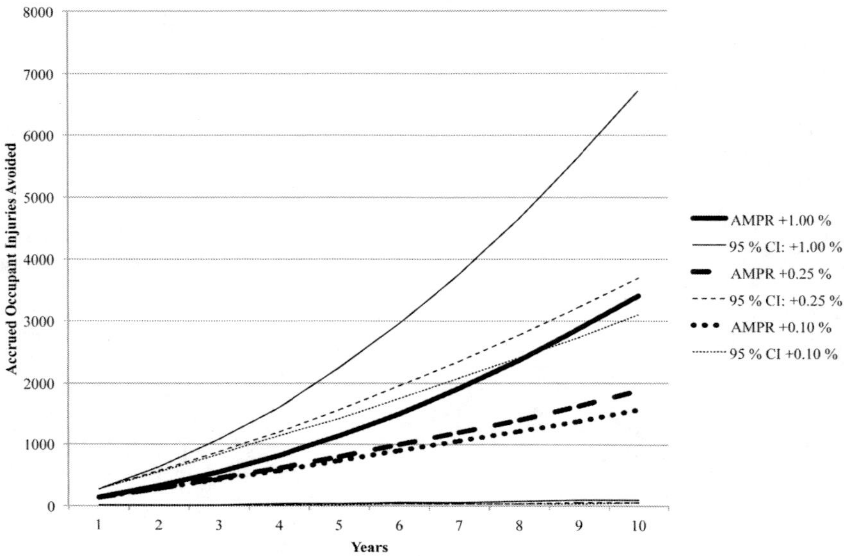

sprinkler requirement is fixed to 10 % (corresponds to the AMPR 0.10 percentage point), 25 % (corresponds to the AMPR 0.25 percentage point), or 100 % (corresponds to the AMPR 1.00 percentage point)[4].

With a 0.10 percentage point annual increase in the housing stock, sprinklers are projected to avert no fewer than 531 occupant fatalities (**Fig. 1**), 49 occupant injuries (**Fig. 2**), and 6010 fire service injuries (**Fig. 3**), over the ten-year period, with the most likely values expected to be 743, 1569, and 9810, respectively. As discussed in footnote 4, this likely approximates the future trend should no more states or local jurisdictions require sprinklers.

With a 0.25 percentage point annual increase in the housing stock, sprinklers are projected to avert no fewer than 635 occupant fatalities (**Fig. 1**), 58 occupant injuries (**Fig. 2**), and 7189 fire service injuries (**Fig. 3**), over the ten-year period, with the most likely values expected to be 889, 1876, and 11 730, respectively.

With a 1.00 percentage point annual increase in the housing stock, the assumed upper limit to sprinkler penetration given recent market turnover, sprinklers are

[4]California and Maryland, states that have adopted the IRC home fire sprinkler requirements, contained 11 % of all single family detached housing units in the U.S. in 2000 [16]. Assuming a 1 % annual turnover in housing in these states, suggests that the 0.10 AMPR depicts current realities. The 1.00 AMPR would depict the upper bound for the Nation.

FIG. 3 Projected number of fire service injuries averted accruing over time, with associated 95 % confidence interval, from residential fire sprinklers, using different annual market penetration rates (AMPR), starting with 3 % of all single family homes sprinklered in Year 1.

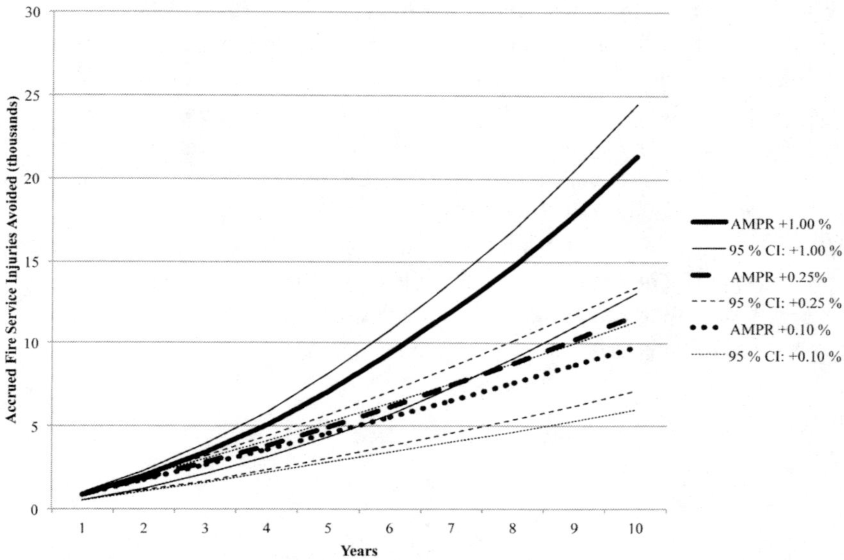

projected to avert no fewer than 1155 occupant fatalities (**Fig. 1**), 106 occupant injuries (**Fig. 2**), and 13 065 fire service injuries (**Fig. 3**), over the ten-year period, with the most likely values expected to be 1616, 3411, and 21 327, respectively.

Conclusion

Residential fire sprinklers offer occupants protection from home fires, as they reduce the likelihood of bodily harm, particularly death. Fires in sprinklered homes also tend to be less injurious to members of the fire service. In an analysis of fire statistics comparing outcomes from sprinklered and non-sprinklered home fires, fires in sprinklered home produce 100 % (95 % confidence interval: ± 29 %) fewer occupant fatalities, 45 % (95 % confidence interval: ± 44 %) fewer occupant injuries, and 87 % (95 % confidence interval: ±34 %) fewer fire service injuries. Yet, 71 % of all fire fatalities occur annually in one- and two- family homes. Currently, fire sprinklers comprise only a small portion of the housing stock of single-family homes. Their scarcity is partly explained by economic considerations and regulatory requirements.

Using ASTM Standard Practices, we demonstrated that fire sprinklers can be economical to homeowners, and that their benefit-cost performance is driven by the (1) physical performance of sprinklers in terms of fatalities avoided, (2) value of a statistical life (VSL), (3) time value of money (discount rate), (4) insurance savings for owners of sprinklered homes, and (5) sprinkler installation and maintenance costs. Only until recently have the benefits been shown to outweigh the costs, specifically when the VSL is high and the discount rate is low [4].

While the number of homes with fire sprinklers is low, it is growing. As state and local jurisdictions adopt building codes, such as the International Residential Code, requiring sprinklers in all new single-family homes, the numbers will increase, although it is unclear how quickly. In this analysis, we evaluated the impact—in terms of occupant fatalities and occupant and firefighter injuries avoided—from continued penetration of sprinklers into the U.S. housing stock. Over the next decade, we predict sprinklers will reduce between 531 to 1616 occupant fatalities, between 49 to 6716 occupant injuries, and between 6010 to 24 521 fire service injuries. The results suggest that the impact of residential fire sprinklers can be significant, but are moderated by the rate of statewide code adoption and the turnover in the housing market.

References

[1] Madrzykowski, D. and Fleming, R. P., 2008, "Residential Sprinkler Systems," *Fire Protection Handbook*, 20th ed., National Fire Protection Association, Quincy, MA, 2008. pp. 16-91–16-106.

[2] National Fire Protection Association, 2014, "One and Two Family Home Fires," http://www.nfpa.org/research/reports-and-statistics/fires-by-property-type/residential/one-and-two-family-home-fires (Last accessed 24 Feb 2014).

[3] National Fire Protection Association, 2014, "Overview of the U.S. Fire Problem," http://www.nfpa.org/research/reports-and-statistics/fires-in-the-us/overall-fire-problem/overview-of-the-us-fire-problem (Last accessed 24 Feb 2014).

[4] Butry, D. T., Brown, M. H., and Fuller, S. K., *Benefit-Cost Analysis of Residential Fire Sprinkler Systems*, National Institute of Standards and Technology, Gaithersburg, MD, 2007.

[5] Hall, J.R., Jr., *U.S. Experience With Sprinklers*, National Fire Protection Association, Quincy, MA, 2013.

[6] Karter, M.J., Jr. and Molis, J. L., *U.S. Firefighter Injuries*, National Fire Protection Association, Quincy, MA, 2013.

[7] Brown, M. H., *Economic Analysis of Residential Fire Sprinkler Systems*, National Institute of Standards and Technology, Gaithersburg, MD, 2005.

[8] ASTM E1074-09: Practice for Measuring Net Benefits and Net Savings for Investments in Buildings and Building Systems, *Annual Book of ASTM Standards*, ASTM International, West Conshohocken, PA, 2012, www.astm.org, DOI:10.1520/E1074-09.

[9] ASTM E917–13: Standard Practice for Measuring Life-Cycle Costs of Buildings and Building Systems, *Annual Book of ASTM Standards*, ASTM International, West Conshohocken, PA, 2012, www.astm.org, DOI:10.1520/E917-05.

[10] ASTM E1369-11: Standard Guide for Selecting Techniques for Treating Uncertainty and Risk in the Economic Evaluation of Buildings and Building Systems, *Annual Book of ASTM Standards*, ASTM International, West Conshohocken, PA, 2012, www.astm.org, DOI:10.1520/E1369-11.

[11] International Code Consortium, 2009, "2009 International Residential Code for One and Two-Family Dwellings," https://archive.org/details/gov.law.icc.irc (Last accessed 28 Feb 2014).

[12] Butry, D. T., "Comparing the Performance of Residential Fire Sprinklers with Other Life Safety Technologies," *Acc. Anal. Prev.*, Vol. 48, 2012, pp. 480–494.

[13] Butry, D. T., "Economic Performance of Residential Fire Sprinkler Systems," *Fire Technol.*, Vol. 45, 2009, pp. 117–143.

[14] ASTM E1121-12: Standard Practice for Measuring Payback for Investments in Buildings and Building Systems, *Annual Book of ASTM Standards*, ASTM International, West Conshohocken, PA, 2012, www.astm.org, DOI:10.1520/E1121-12.

[15] Housing and Urban Development, 2014, "*AHS Components of Inventory Change: 2007-2009*," http://www.huduser.org/portal/datasets/cinch/cinch09/cinch07-09.pdf (Last accessed 26 Feb 2014).

[16] Census Bureau, 2014, "Historical Census of Housing Tables," https://www.census.gov/hhes/www/housing/census/historic/units.html (Last accessed 24 Feb 2014).

STP 1586, 2014 / available online at www.astm.org / doi: 10.1520/STP158620140036

J. D. Kneifel[1]

Life-Cycle Cost Implications of More Stringent State Energy Codes

Reference

Kneifel, J. D., "Life-Cycle Cost Implications of More Stringent State Energy Codes," *Ideas to Impact: How Building Economic Standards Keep You on Track*, STP 1586, Robert E. Chapman and Michael N. Goodkind, Eds., pp. 164–183, doi:10.1520/STP158620140036, ASTM International, West Conshohocken, PA 2014.[2]

ABSTRACT

Energy efficiency requirements in energy codes vary across states, with states having adopted energy codes ranging across editions of the industry consensus standard (ASHRAE 90.1). Some states do not have a code requirement for energy efficiency, leaving it up to the locality or jurisdiction to set its own requirements. This paper uses ASTM building economic standards (E917-05(2010) and E1074-09) to estimate the impacts that the adoption of more stringent energy codes for commercial buildings would have on building life-cycle costs. The results are based on analysis of the Building Industry Reporting and Design for Sustainability (BIRDS) database. For this study, the performance of buildings designed to meet current state energy codes is compared to their performance when meeting a "Low Energy Case" (LEC) building design based on ASHRAE 189.1-2009, which increases energy efficiency beyond the ASHRAE 90.1-2007 design, to determine whether more stringent energy standard editions are life-cycle cost-effective in reducing energy consumption. The approach is described in detail for a single city and building type (Knoxville, TN). Using the same approach, the new savings for each building type in all cities for Tennessee are calculated. The estimated average savings for each of the building types are aggregated using state-level new commercial building construction data to calculate the magnitude of the net savings

---

Manuscript received April 1, 2014; accepted for publication April 25, 2014; published online July 14, 2014.

[1]Applied Economics Office, Engineering Laboratory, National Institute of Standards and Technology, Gaithersburg, MD 20899.

[2]ASTM Symposium on *Ideas to Impact: How Building Economic Standards Keep You on Track* on October 5, 2014 in New Orleans, LA.

(80.8 GWh annually and $\$28.1 \times 10^6$ in life-cycle costs) that Tennessee may realize if it were to adopt the LEC design as its state energy code. These state-level estimates are further aggregated to the national level, estimating the potential total impact from nationwide adoption of the LEC design to be 34 441 GWh and $1.0 billion for 1 years' worth of construction for a 10-year study period.

Keywords
building economics, economic analysis, energy efficiency, life-cycle costing

Introduction

Energy efficiency requirements in energy codes vary across states, with states having adopted energy codes ranging across editions of the industry consensus standard (ASHRAE 90.1 [1–4]). Some states do not have a code requirement for energy efficiency, leaving it up to the locality or jurisdiction to set its own requirements. This paper uses ASTM building economic standards (ASTM E917-05(2010) [5] and ASTM E1074-09 [6]) to estimate the impacts that the adoption of more stringent energy codes for commercial buildings would have on building life-cycle costs.

The results of this paper are based on analysis of the Building Industry Reporting and Design for Sustainability (BIRDS) database, which includes life-cycle cost analysis based on 12 540 whole building energy simulations covering 11 building types in 228 cities across all U.S. states for 9 study period lengths. For this study, the performance of buildings designed to meet current state energy codes is compared to their performance when meeting a "Low Energy Case" (LEC) building design based on ASHRAE 189.1-2009 [7], which increases energy efficiency beyond the ASHRAE 90.1-2007 [4] design, to determine whether more stringent energy standard editions are life-cycle cost-effective in reducing energy consumption.

The approach is described in detail for a single city and building type. Using the same approach, the new savings for each building type in all cities for a state are calculated. The estimated average savings for each of the building types are aggregated using state-level new commercial building construction data to calculate the magnitude of the net savings that a state may realize if it were to adopt a more energy efficient standard edition as its state energy code. These state-level estimates are further aggregated to the national level to estimate the potential total impact from nationwide adoption of more stringent energy codes.

Study Design

The building characteristics in **Table 1** describe the 11 building types used in this study, which include 2 dormitories, 2 apartment buildings, a hotel, 3 office

TABLE 1 Building characteristics.

Building Type	Building Abbreviaton	Floors	Floor Height m (ft)	Wall	Roof[a]	Percent Glazing	Building Size m^2 (ft^2)	Occupancy Type	U.S. Floor Space (%)
Dormitory	DORMI04	4	3.66 (12)	Mass	IEAD	20 %	3097 (33 333)	Lodging	7.1 %
Dormitory	DORMI06	6	3.66 (12)	Steel	IEAD	20 %	7897 (85 000)		
Hotel	HOTEL15	15	3.05 (10)	Steel	IEAD	100 %	41 806 (450 000)		
Apartment	APART04	4	3.05 (10)	Mass	IEAD	12 %	2787 (30 000)		
Apartment	APART06	6	3.15 (10)	Steel	IEAD	14 %	5574 (60 000)		
School, High	HIGHS02	2	4.57 (15)	Mass	IEAD	25 %	12 077 (130 000)	Education	13.8 %
Office	OFFIC03	3	3.66 (12)	Mass	IEAD	20 %	1858 (20 000)	Office	17.0 %
Office	OFFIC08	8	3.66 (12)	Mass	IEAD	20 %	7432 (80 000)		
Office	OFFIC16	16	3.05 (10)	Steel	IEAD	100 %	24 155 (260 000)		
Retail Store	RETAIL1	1	4.27 (14)	Mass	IEAD	10 %	743 (8000)	Mercantile[b]	6.0 %
Restaurant	RSTRNT1	1	3.66 (12)	Wood	IEAD	30 %	465 (5000)	Food Service	2.3 %

[a]IEAD = Insulation entirely above deck.
[b]Only includes non–mall floor area.

FIG. 1 State commercial energy codes.

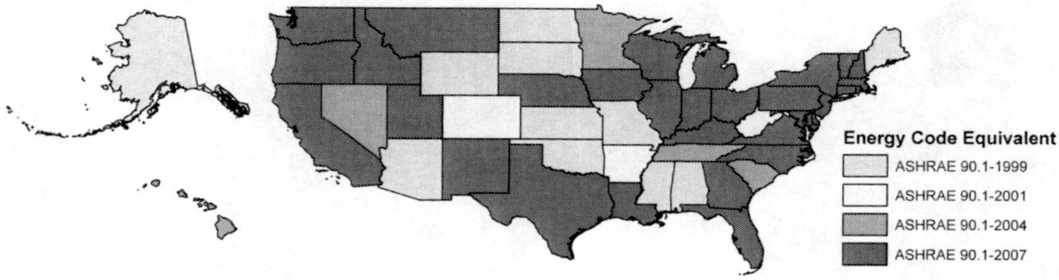

buildings, a school, a retail store, and a restaurant. The building types were selected based on a combination of factors, including fraction of building stock represented, variation in building characteristics, and ease of simulation design. These building types represent 46 % of the existing U.S. commercial building stock floor space[3]. The prototype buildings range in size from 465 m^2 (5000 ft^2) to 41 806 m^2 (450 000 ft^2)[4]. The building abbreviations defined in **Table 1** are used to represent the building types in tables throughout this study.

Current state energy codes are based on different editions of the International Energy Conservation Code (IECC) or ASHRAE 90.1 Standard, which have requirements that vary based on a building's characteristics and the climate zone of the building location. For this study, the prescriptive requirements of the ASHRAE 90.1 Standard-equivalent design are used to meet current state energy codes. States that have not yet adopted a state energy code are assumed to meet ASHRAE 90.1-1999 building energy efficiency requirements. A "Low Energy Case" design based on ASHRAE 189.1-2009, which goes beyond ASHRAE 90.1-2007, is the building design alternative.

Figure 1 shows that commercial building energy codes as of December 2011 vary by state from no code (assumed to be ASHRAE 90.1-1999) to ASHRAE 90.1-2007[5]. In a few instances, local jurisdictions have adopted energy standard editions that are more stringent than the state energy codes.[6,7]

The 228 cities and ASHRAE climate zones for the U.S. are seen in **Fig. 2**. The large selection of cities allows for analysis of variation in results within a state and across climate zones for the most populated cities in the country.

---

[3]Estimate is based on the Commercial Building Energy Consumption Survey (CBECS) database [8].
[4]The whole building energy simulations are created using the EnergyPlus Example File Generator [9].
[5]Since the publication of Kneifel [10,11], the BIRDS database has been updated to include subsequent changes in state energy codes through December 2011.
[6]Local and jurisdictional requirements are obtained from the Database of State Incentives for Renewables and Efficiency (DSIRE) [12]. State energy code requirements targeting only public buildings and green standards are ignored in this study.
[7]Figure was obtained from the DOE Building Technologies Program in December 2011 [13].

FIG. 2 Cities and ASHRAE climate zones.

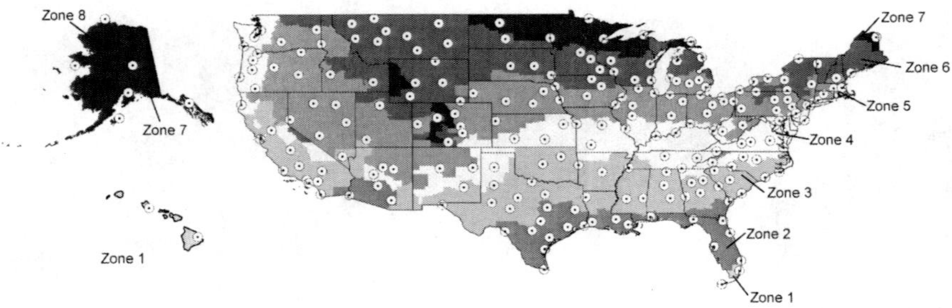

Analysis Approach

The analysis in this report compares energy and life-cycle cost performance of the status quo state energy codes to the more stringent LEC design alternative. For any state without a state energy code, ASHRAE 90.1-1999 [1] is assumed to be the baseline because it represents minimum energy–related industry practices. It is assumed that the building maintains its energy efficiency performance throughout the study period, resulting in energy consumption remaining constant over the entire study period. This assumption is reasonable given the maintenance, repair, and replacement costs included in the analysis to ensure the building and its equipment perform as expected.

Two metrics are used to analyze the annual energy performance: simple percentage savings and total net energy savings. The average percentage energy use savings are calculated by taking the simple average of the percentage savings for each location–building type combination in a city.

ASTM E917–05(2010) [5] (Standard Practice for Measuring Life-Cycle Costs of Buildings and Building Systems) is used to calculate each building design's life--cycle costs (LCCs). LCCs take into account all relevant costs throughout the chosen study period, including construction costs, maintenance, repair, and replacement costs, energy costs, and residual values. A cost's present value (PV) is calculated by discounting its nominal value into today's dollars based on the year the cost occurs and the assumed discount rate, which is assumed to be 3 % for this study[8]. LCC of buildings typically compares the costs for a baseline building design to the costs for alternative, more energy–efficient building designs to determine if future operational savings justify higher initial investments. For this study, the LEC design is compared to the baseline state energy code compliant design to determine the changes in life-cycle costs. The metric used to analyze the economic performance

[8]The 2013 discount rate defined for federal energy efficiency projects.

for a building is Net Savings, which is the difference between the alternative design and base case's LCCs as defined in ASTM E1074-09 [6] (Standard Practice for Measuring Net Benefits and Net Savings for Investments in Buildings and Building Systems).

The estimated average changes in total energy use and life–cycle costs for each of the building types is combined with state-level new commercial building construction data to calculate the magnitude of the available total savings a state may realize if it were to adopt a more energy efficient standard as its state energy code. Using this approach to estimate the total impacts for a state requires the assumption that the savings realized by the BIRDS building prototypes are representative of new commercial building construction as a whole.

It is necessary to assume a particular study period length to generate results. Although the annual energy use savings is assumed to be the same across study period lengths, the life-cycle costs vary with the study period length because costs vary year-over-year. A 10-year study period is used for the analysis.

Data

The cost of data collected to estimate life-cycle costs for the BIRDS database originates from multiple sources, including RS Means databases [14], Whitestone [15], and the U.S. Energy Information Administration (EIA) [16,17][9]. Costs are grouped into two categories, first costs that include initial building construction costs and future costs that include operational energy costs, maintenance, repair, and replacement (MRR) costs, and building residual value.

Building construction costs are obtained from the RS Means CostWorks online databases [14]. The costs of a prototypical building are estimated by the RS Means CostWorks Square Foot Estimator to obtain the default costs of each component within each separate building type. The RS Means default building is the baseline used to create a building that is compliant with both the state energy code and the LEC design. The RS Means default buildings are adapted to match the five prototype building standard designs by using the RS Means CostWorks Cost Books databases.

Five components—roof insulation, wall insulation, windows, lighting, and heating, ventilation, and air conditioning (HVAC) efficiency—are changed to make the prototypical designs code compliant. The LEC design increases the thermal efficiency of insulation and windows beyond ASHRAE 90.1-2007, further reduces the lighting power density, and adds daylighting and window overhangs[10]. The lighting density of the lighting system is decreased by first increasing the efficiency of the

---

[9]See Kneifel [11] for additional details of the cost data used in the BIRDS database.
[10]Overhang cost source is Winiarski et al. [18]

lighting system and then decreasing the number of fixtures in the lighting system[11]. Daylighting is included for all building types and climate zones. Overhangs are placed on the east, west, and south sides of the building for each floor in climate zone 1 through climate zone 5 because these warmer climates are the zones that benefit from blocking solar radiation.

The HVAC system size varies across building designs because changing the thermal characteristics of the building envelope alters the heating and cooling loads of the building. The EnergyPlus [21] whole building energy simulations "autosize" the HVAC system to determine the appropriate system size to efficiently maintain the thermal comfort while dealing with ventilation requirements. For each building design, the HVAC cost for the default HVAC system is replaced with the cost of the "autosized" HVAC system. An HVAC efficiency cost multiplier is used to adjust the HVAC costs in accordance with the standard efficiency requirements.

Construction costs for a building in each location are estimated by summing the baseline costs for the RS Means default building and the changes in costs required to meet the alternative prototype designs, each of which is adjusted with the 2009 RS Means CostWorks City Indexes to control for local material and labor price variations as shown in the following equation where "Index" is the city cost index, "*j*" is the building–location combination, and "*c*" is building component:

$$Construction\ Costs_j = Costs_{\mathrm{Baseline}} * Index_j + \sum_{c}^{C} \Delta Cost_{jc} * Index_{jc} \qquad (1)$$

The indexed construction costs are then multiplied by a contractor "mark-up" rate, 25 %, and architectural fees rate, 7 %, which are added to the construction costs to estimate the building's "first costs" of construction for the prototype buildings[12] as shown in the following equation:

$$First\ Costs_j = Construction\ Costs_j * 125\ \% * 107\ \% \qquad (2)$$

Future costs include maintenance, repair and replacement costs, residual value, and operational energy costs. Component and building lifetimes and component repair requirements are based on data from Whitestone [15]. Building component maintenance, repair, and replacement (MRR) rates are from Kneifel [22] and Kneifel [23]. Insulation and windows are assumed to have a lifespan greater than 40 years and have no maintenance requirements. Insulation is assumed to have no repair costs. Windows have an assumed annual repair cost equal to replacing 1 % of all window

---

[11]First, incandescent lighting is replaced with compact fluorescent lighting while typical T-12 fluorescent tube lighting is replaced with more efficient T-8 fluorescent tube lighting to decrease the lighting density of the lighting system. Second, the number of fixtures is reduced to meet the remainder of the required reduction in watts per unit of floor area. Increasing the efficiency of the lighting increases the costs of construction. The first approach increases first costs while the second approach decreases first costs for the lighting system. This approach is based on Belzer et al. [19] and Halverson et al. [20].

[12]These rates are the default values used by the RSMeans Square Foot Estimator [14].

panes, with costs that vary depending on the required window specifications [14]. The heating and cooling units have different lifespans and repair rates based on climate, ranging from 4 years to 33 years for repairs and 13 years to 50 years for replacements [15].

MRR cost data are collected from two sources. The total maintenance and repair costs per square foot of conditioned floor area (minus the HVAC maintenance and repair costs) represent the baseline MRR costs per unit of floor area, which occur for a building type regardless of the energy efficiency measures incorporated into the design. These data are collected from Whitestone [15], which reports average maintenance and repair costs per unit of floor area by building component for each year of service life for each building type. The building types in Whitestone do not match exactly to the 11 building types selected for this study, so the most comparable profile is selected.

RS Means CostWorks [14] is the source of MRR costs for the individual components for which MRR costs change across alternative building designs (HVAC system, lighting system, and windows). Lighting systems, including daylighting controls for the LEC design, are assumed to be replaced every 20 years. The HVAC system size varies based on the thermal performance of the alternative building design, which results in varying MRR costs because smaller systems are relatively cheaper to maintain, repair, and replace.

Future MRR costs are discounted to equivalent present values using the single present value (SPV) factors for future non–fuel costs reported in Rushing and Lippiatt [24], which are calculated using the USDOE's 2008 real discount rate for energy conservation projects (3 %) as shown in the following equation where "*Index*" is the city cost index, "*j*" is the building–location combination, "*t*" is the study period, "*Cost*" is MRR costs for a given year, and "*SPV*" is single present value:

$$MRR\ Costs_j = Index_j * \sum_{i}^{t}(Cost_{ji} * SPV_i) \tag{3}$$

A building's residual value is its value at the end of the study period. It is estimated in two parts, original components for the building excluding components potentially replaced during the study period and building components potentially replaced during the study period (HVAC system and lighting system) based on the approach defined in Fuller and Petersen [25]. The residual value of the original building components is assumed to be equal to the building's first cost minus any components replaced over the study period (*First Costs*$_j$) multiplied by the ratio of the study period to the service life of the building ($(L_B - t)/L_B$), and discounted from the end of the study period.

Two components may be replaced during the study period, the lighting and HVAC systems. Residual values for these components are computed for each location in a similar manner to the building residual value. The remaining "life" of the component is determined by taking its service life minus the number of years since its last installation, whether it occurred during building construction or replacement. The ratio of remaining life to service life ($t\ mod\ L_{jc}/L_{jc}$) is multiplied by the installed cost of the lighting and HVAC systems ($Replace\ Costs_c$), and discounted from the end of the study period. The lighting system service life is 20 years while the HVAC system service life varies by location based on Towers et al. [15].

The following equation shows the combined residual value for the original and potentially replaced building components where "*RV*" is residual value, "*j*" is the building–location combination, "*t*" is the study period, "*L*" is useable life, "*B*" is building, "*c*" is building component potentially replaced, "*C*" is the total number of building components potentially replaced, and "*SPV*" is single present value:

$$RV_j = \left(First\ Costs_j * \frac{L_B - t}{L_B} * SPV_t\right) + \sum_{c}^{C}\left(Replace\ Costs_{jc} * \frac{t\ mod\ L_{jc}}{L_{jc}} * SPV_t\right) \quad (4)$$

Annual energy costs are estimated by multiplying annual electricity and natural gas use predicted by the whole building energy simulation by the average state retail commercial electricity and natural gas prices, respectively. Average state commercial electricity and natural gas prices for 2009 are collected from the Energy Information Administration (EIA) Electric Power Annual State Data Tables [16] and Natural Gas Navigator [17], respectively. The electricity and natural gas prices are assumed to escalate over time according to EIA forecasts from 2009 to 2039. These forecasts are embodied in the Federal Energy Management Program (FEMP) Uniform Present Value Discount Factors for energy price estimates (UPV^*) reported in Rushing and Lippiatt [24][13]. The UPV^* values are used to discount future energy costs to equivalent present values as shown in the following equation where "*j*" is the building–location combination, "*t*" is the study period, "*e*" is electricity, "*g*" is natural gas, "*P*" is energy price, "*Use*" is annual energy consumption, and "UPV^*" is the modified uniform present value. The term modified is used because the UPV formula includes a price escalation rate. The discount factors vary by Census region, building sector, and fuel type.

$$Energy\ Costs_j = P_{je} * Use_{je} * UPV^*_{tg} + P_{jg} * Use_{jg} * UPV^*_{tg} \quad (5)$$

Life–cycle costs are the sum of the first costs and future costs as shown in the following equation where "*j*" is the building–location combination:

$$LCC_j = First\ Costs_j + MRR\ Costs_j - RV_j + Energy\ Costs_j \quad (6)$$

Aggregating the savings for newly constructed commercial buildings to the state and national levels requires new construction data for each building type within each state. This study uses the commercial building weighting factors reported

[13]The escalation rates for years 31 to 40 are assumed to be the same as for year 30.

in Jarnagin and Bandyopadhyay [26] to estimate the total energy use savings and life-cycle cost savings resulting from adopting newer energy standard editions for each state. Jarnagin and Bandyopadhyay [26] use two databases to generate the commercial building weighting factors: the 2003 Commercial Buildings Energy Consumption Survey (CBECS) [8] and a McGraw–Hill construction dataset [27].

The McGraw–Hill dataset includes data for all new commercial buildings and additions, over 254 000 records and 761.8×10^6 m^2 (8.2×10^9 ft^2) of new construction, for 2003–2007. The data are more detailed than the CBECS data, and includes year of construction and location. CBECS includes 14 building type categories: education, food sales, food service, health care, lodging, mercantile, office, public assembly, public order and safety, religious worship, service, warehouse and storage, other, and vacant. Each category includes up to 12 subcategories. Jarnagin and Bandyopadhyay [26] maps the more detailed McGraw–Hill dataset to the CBECS categories and subcategories. The McGraw–Hill dataset [27] is aggregated at the CBECS category–level. For this study, a prototype building is assumed to represent its entire CBECS category, which implies the prototypes together represent 56.8 % of the new commercial building stock in the U.S. from 2003 to 2007.

Results

This section analyzes a subset of the BIRDS database results for net savings in energy use and life-cycle costs at the building, city, state, and national level. The analysis starts at the building level considering the most common building type, a low–rise office building, in Knoxville, TN. The same approach is replicated for all eleven building types for five cities in Tennessee (all Tennessee cities in the BIRDS database). The city–level results are then aggregated to the state level using new construction data for Tennessee. Finally, the state level results are aggregated to the national level to estimate the net savings from a nationwide adoption of the LEC design.

THREE-STORY OFFICE BUILDING IN KNOXVILLE, TN

The analysis in this study begins by looking at a case study comparing the energy and economic performance of adopting the LEC design as the energy code for a low-rise office building in Knoxville, TN relative to the current state energy code, which is based on ASHRAE 90.1–2004. The building prototype is a 3–story, 1858 m^2 (20 000 ft^2) building.

Constructing an 1858 m^2 (20 000 ft^2) office building to meet the LEC design leads to a decrease in both energy use and life–cycle costs. As shown in **Table 2**, annual energy consumption for the building is decreased by 51 383 kWh (25.8 %) while life–cycle costs are decreased by $41 350 for a 10–year study period. The additional initial investment costs required to increase the thermal performance of the building envelope is more than offset by the energy cost savings over the 10–year study period.

TABLE 2 Net savings from adoption of the LEC design—Knoxville, TN—10-year study period.

	Net Savings	
Metric	**Annual Energy Use**	**Life-Cycle Costs**
Percent	25.8 %	3.1 %
Total	51 383 kWh	$41 350

Table 3 shows this approach applied to all 11 building types for a 10-year study period. All 11 building types realize energy use savings, ranging from 20 014 to 552 267 kWh (10.5 %–34.0 %). The net savings in life-cycle costs varies across building types, with net savings ranging from −$271 609 for the hotel to $169 671

TABLE 3 Change from adoption of the LEC design — Knoxville, TN — 10-Year study period: (a) metric units (b) customary units.

(a)	Building Type	Floor Area (m^2)	Δ Annual Energy Use (kWh) Total	Percent	Per m^2	Δ Life-Cycle Costs Total	Percent	Per m^2
	APART04	2787	−61 678	−16.2	−22.6	−$2197	−0.1	−$0.75
	APART06	5574	−133 625	−17.7	−23.7	−$13 684	−0.3	−$2.48
	DORMI04	3097	−56 975	−18.1	−18.3	−$15 937	−0.6	−$5.17
	DORMI06	7897	−175 100	−16.0	−23.7	−$32 649	−0.5	−$4.41
	HOTEL15	41 806	−552 267	−10.5	−12.9	$271 609	0.8	$6.46
	HIGHS02	12 077	−289 669	−13.6	−23.7	−$164 872	−1.9	−$13.67
	OFFIC03	1858	−51 383	−25.8	−28.0	−$41 350	−3.1	−$22.28
	OFFIC08	7432	−181 067	−23.7	−24.8	−$169 671	−3.1	−$22.82
	OFFIC16	24 155	−404 503	−13.0	−17.2	$123 855	0.8	$5.17
	RETAIL1	743	−20 014	−23.5	−26.9	−$1738	−0.4	−$2.37
	RSTRNT1	465	−25 006	−34.0	−53.8	−$45 780	−8.3	−$98.60

(b)	Building Type	Floor Area (ft^2)	Δ Annual Energy Use (kWh) Total	Percent	Per ft^2	Δ Life-Cycle Costs Total	Percent	Per ft^2
	APART04	30 000	−61 678	−16.2	−2.1	−$2197	−0.1	−$0.07
	APART06	60 000	−133 625	−17.7	−2.2	−$13 684	−0.3	−$0.23
	DORMI04	33 333	−56 975	−18.1	−1.7	−$15 937	−0.6	−$0.48
	DORMI06	80 000	−175 100	−16.0	−2.2	−$32 649	−0.5	−$0.41
	HOTEL15	450 000	−552 267	−10.5	−1.2	$271 609	0.8	$0.60
	HIGHS02	130 000	−289 669	−13.6	−2.2	−$164 872	−1.9	−$1.27
	OFFIC03	20 000	−51 383	−25.8	−2.6	−$41 350	−3.1	−$2.07
	OFFIC08	80 000	−181 067	−23.7	−2.3	−$169 671	−3.1	−$2.12
	OFFIC16	260 000	−404 503	−13.0	−1.6	$123 855	0.8	$0.48
	RETAIL1	8000	−20 014	−23.5	−2.5	−$1738	−0.4	−$0.22
	RSTRNT1	5000	−25 006	−34.0	−5.0	−$45 780	−8.3	−$9.16

for the mid-rise office building. Depending on the building type, the net savings can vary significantly from $-\$6.46/m^2$ to $-\$98.60/m^2$ ($-\$0.60/ft^2$ to $9.16/ft^2$). Two building types (high-rise hotel and office building) realize an increase in life-cycle costs.

TENNESSEE

The approach used in calculating the net savings per unit of floor area for buildings in Knoxville are used to estimate the impacts for four other cities in Tennessee (Bristol, Chattanooga, Memphis, and Nashville). **Table 4** shows the changes per m^2 (ft^2) of floor area for energy use and life-cycle costs from adoption of the LEC design relative to the ASHRAE 90.1-2004 [3] design. The simple average percentage reduction in energy use for the eleven building types varies from 17.8 to 19.4 % (not shown) across the five cities, which is an average reduction of 23.7–25.8 kWh/m^2 (2.2–2.4 kWh/ft^2). Average net savings in life-cycle costs for all eleven building types ranges from $\$12.38/m^2$ to $\$15.39/m^2$ ($\$1.15/ft^2$–$\$1.43/ft^2$) on average across the five cities. Similar to Knoxville, the net savings per unit of floor area varies across building types for each city in Tennessee, with the average across cities ranging from $-\$7.10/m^2$ to $\$96.01/m^2$ ($-\$0.66/ft^2$ to $\$8.92/ft^2$) for a given building type.

In order to determine the aggregate net savings in energy use and life-cycle costs the state of Tennessee could realize from the adoption of the LEC design, it is necessary to match the net savings per unit of floor area to the amount of new construction in Tennessee for each building type. In this case, the available new construction data is at the state level, which requires the use of the average change per unit of floor area for all cities for Tennessee in the database. Consider the low-rise office building in Tennessee. Average annual statewide construction of low-rise office buildings totaled 0.17×10^6 m^2 (1.85×10^6 ft^2) for 2003–2007. The average net savings in energy use across the five cities in Tennessee for the 3-story office building is 28.0 kWh/m^2 (2.6 kWh/ft^2) annually while average net savings in life-cycle costs for the 10-year study period is $\$18.94/m^2$ ($\$1.76/ft^2$). As a result, the aggregate net savings from new low-rise office building construction for the state of Tennessee totals:

$$\textit{Total Energy Savings} = \frac{2.6\,\frac{\text{kWh}}{\text{years}}}{\text{ft}^2} * 10 \text{ years} * 1.85 \times 10^6 \text{ ft}^2 = 48.10 \times 10^6 \text{ kWh} \tag{7}$$

$$\textit{Total Life} - \textit{Cycle Cost Savings} = \$1.76 * 1.85 \times 10^6 \text{ ft}^2 = \$3.25 \times 10^6 \tag{8}$$

The aggregate net savings for nine of the eleven building prototypes is shown in **Table 5** (the new construction data does not include dormitories as a category as is excluded from the remaining analysis). In total, new construction represented by the nine building prototypes leads to net savings in annual energy use of 47.2×10^6 kWh and net savings in life-cycle costs over the 10-year study period of $\$16.4 \times 10^6$. The nine prototypes represent 58.4 % of total new construction.

TABLE 4 Change in annual energy use and life-cycle costs from adoption of the LEC design by city—Tennessee, 10-year study period: (a) per m^2 (b) per ft^2.

(a)

Building Type	Memphis		Bristol		Chattanooga		Knoxville		Nashville		Average	
	kWh	PV$	kWh	PV$	kWh	PV$	kWh	PV$	kWh	PV$	kWh	PV$
APART04	−23.7	−$2.48	−21.5	−$1.51	−22.6	$0.32	−22.6	−$0.75	−23.7	−$0.54	−22.6	−$0.97
APART06	−24.8	−$3.55	−22.6	−$2.80	−24.8	−$1.51	−23.7	−$2.48	−24.8	−$1.61	−23.7	−$2.37
DORMI04	−16.1	−$13.89	−18.3	−$20.77	−18.3	−$3.44	−18.3	−$5.17	−19.4	−$5.70	−18.3	−$9.80
DORMI06	−29.1	−$8.93	−21.5	−$4.63	−23.7	−$3.44	−23.7	−$4.41	−23.7	−$3.66	−24.8	−$5.06
HOTEL15	−9.7	$7.32	−12.9	$5.81	−12.9	$7.86	−12.9	$6.46	−12.9	$7.86	−11.8	$7.10
HIGHS02	−21.5	−$10.01	−21.5	−$13.24	−23.7	−$12.16	−23.7	−$13.67	−24.8	−$13.78	−22.6	−$12.59
OFFIC03	−24.8	−$9.58	−26.9	−$29.60	−28.0	−$14.64	−28.0	−$22.28	−30.1	−$18.51	−28.0	−$18.94
OFFIC08	−22.6	−$17.65	−24.8	−$30.35	−24.8	−$20.45	−24.8	−$22.82	−25.8	−$24.76	−24.8	−$23.25
OFFIC16	−14.0	$6.35	−16.1	$4.41	−17.2	$6.35	−17.2	$5.17	−17.2	$6.67	−16.1	$5.81
RETAIL1	−23.7	−$0.75	−29.1	−$4.31	−28.0	−$0.54	−26.9	−$2.37	−29.1	−$3.98	−26.9	−$2.37
RSTRNT1	−47.4	−$109.15	−53.8	−$66.31	−56.0	−$94.72	−53.8	−$98.60	−54.9	−$111.41	−52.7	−$96.01
Average	−23.7	−$14.75	−24.8	−$14.85	−25.8	−$12.38	−24.8	−$14.64	−25.8	−$15.39	−24.8	−$14.42

(b)

Building Type	Memphis		Bristol		Chattanooga		Knoxville		Nashville		Average	
	kWh	PV$	kWh	PV$	kWh	PV$	kWh	PV$	kWh	PV$	kWh	PV$
APART04	−2.2	−$0.23	−2.0	−$0.14	−2.1	$0.03	−2.1	−$0.07	−2.2	−$0.05	−2.1	−$0.09
APART06	−2.3	−$0.33	−2.1	−$0.26	−2.3	−$0.14	−2.2	−$0.23	−2.3	−$0.15	−2.2	−$0.22
DORMI04	−1.5	−$1.29	−1.7	−$1.93	−1.7	−$0.32	−1.7	−$0.48	−1.8	−$0.53	−1.7	−$0.91
DORMI06	−2.7	−$0.83	−2.0	−$0.43	−2.2	−$0.32	−2.2	−$0.41	−2.2	−$0.34	−2.3	−$0.47
HOTEL15	−0.9	$0.68	−1.2	$0.54	−1.2	$0.73	−1.2	$0.60	−1.2	$0.73	−1.1	$0.66
HIGHS02	−2.0	−$0.93	−2.0	−$1.23	−2.2	−$1.13	−2.2	−$1.27	−2.3	−$1.28	−2.1	−$1.17
OFFIC03	−2.3	−$0.89	−2.5	−$2.75	−2.6	−$1.36	−2.6	−$2.07	−2.8	−$1.72	−2.6	−$1.76
OFFIC08	−2.1	−$1.64	−2.3	−$2.82	−2.3	−$1.90	−2.3	−$2.12	−2.4	−$2.30	−2.3	−$2.16
OFFIC16	−1.3	$0.59	−1.5	$0.41	−1.6	$0.59	−1.6	$0.48	−1.6	$0.62	−1.5	$0.54
RETAIL1	−2.2	−$0.07	−2.7	−$0.40	−2.6	−$0.05	−2.5	−$0.22	−2.7	−$0.37	−2.5	−$0.22
RSTRNT1	−4.4	−$10.14	−5.0	−$6.16	−5.2	−$8.80	−5.0	−$9.16	−5.1	−$10.35	−4.9	−$8.92
Average	−2.2	−$1.37	−2.3	−$1.38	−2.4	−$1.15	−2.3	−$1.36	−2.4	−$1.43	−2.3	−$1.34

TABLE 5 Statewide change in annual energy use for one year of construction, Tennessee.

Building Type	Subcategory Weighting	m² (1000 s)	ft² (1000 s)	State Level Impacts: Annual Change in Energy Use kWh	kBtu	Change in LCCs PV$
APART04	44.9 %	89	955	−2 010 540	−6 864 847	−$87 633
APART06	55.1 %	109	1169	−2 615 546	−8 930 598	−$260 597
HOTEL15	100.0 %	136	1469	−1 669 894	−5 701 738	$967 430
HIGHS02	100.0 %	362	3895	−8 372 078	−28 585 875	−$4 548 170
OFFIC03	37.4 %	172	1850	−4 739 500	−16 182 692	−$3 248 330
OFFIC08	40.4 %	185	1995	−4 526 951	−15 456 957	−$4 305 144
OFFIC16	22.2 %	102	1098	−1 645 100	−5 617 078	$589 230
RETAIL1	100.0 %	716	7710	−19 482 585	−66 521 922	−$1 703 531
RSTRNT1	100.0 %	40	429	−2 115 887	−7 224 547	−$3 827 052
Total - Prot. (58.4 % of Total)		1911	20 571	−47 178 081	−161 086 253	−$16 423 797
Total-Entire State		3272	35 219	−80 784 385	−275 832 625	−$28 122 940

Note: Dormitories are excluded because no such floor area category is reported in the construction data.

Assuming the net savings realized by the nine prototypes is representative of the net savings that would be realized by all new construction, aggregate net savings for all new construction totals 80.8 GWh annually and $\$28.1 \times 10^6$ over the 10-year study period.

NATIONWIDE

The statewide estimates for energy use and life-cycle costs described for Tennessee have been replicated for all 50 states. As would be expected, the net savings in both energy use and life-cycle costs vary across states due to two factors: the amount of new construction in a state and the current energy code requirements for a state.

Figure 3[14] shows the amount of new floor area (1000 m^2) constructed annually for each of the 50 states from 2003 to 2007. States with a greater amount of new floor area constructed (California, Florida, and Texas) are expected to realize greater reductions in energy use and life-cycle costs.

Figure 4 shows the total net savings in energy use and life-cycle costs from adopting the LEC design as the commercial building energy code for all 50 states for a 10-year study period. The states in hatch marks are the states with either no code or ASHRAE 90.1-2001 as its energy code[15]. In general, the greater the amount of new floor area, the greater the reductions in energy use. California, Florida, and Texas realize the greatest reductions in energy use (1900–3800 GWh). In

[14] $1\ m^2 = 10.76391\ ft^2$.

[15] As of December 2011.

FIG. 3 Average annual new floor area by state.

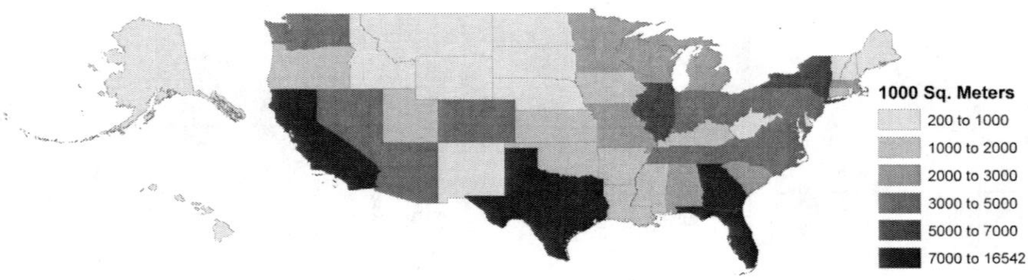

comparison, states with the smallest amount of new floor area (Alaska, Delaware, Monata, Rhode Island, Vermot, and Wyoming) realize savings of less than 100 GWh. There is a strong correlation between total savings in energy use and total net savings in life-cycle costs. California, Florida, and Texas realize the greatest total life-cycle cost savings. The relative total reductions in life-cycle costs vary from total reductions in energy use to some extent, particularly for states with no energy code. Only two states realize an increase in life-cycle costs (Oregon and Washington), and these increases are minimal when compared to the net savings realized by all other states.

Since the amount of new construction is the primary driver of total net savings realized by a state, it is necessary to control for this factor to determine the impacts that a state's current energy code requirements have on net savings. **Figure 5** shows

FIG. 4 Total Net savings in energy use and life-cycle costs by state, 10-year.

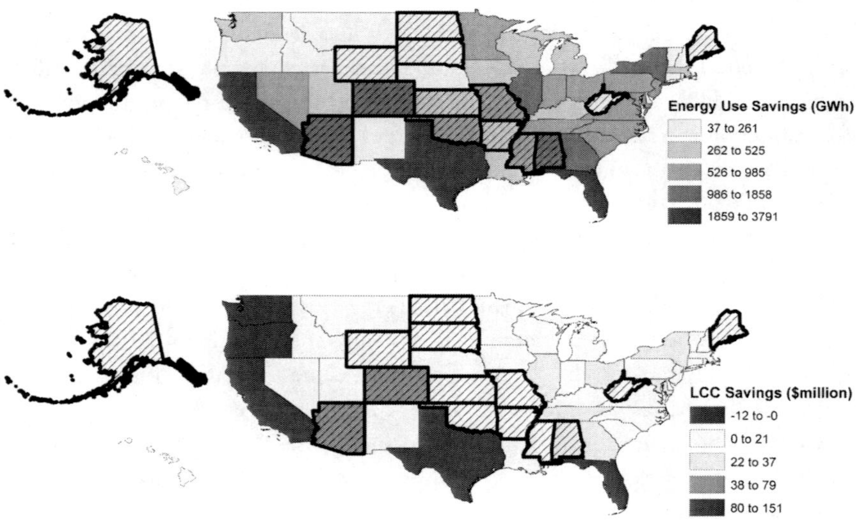

FIG. 5 Net savings in energy use and life-cycle costs per unit of floor area by state, 10-year.

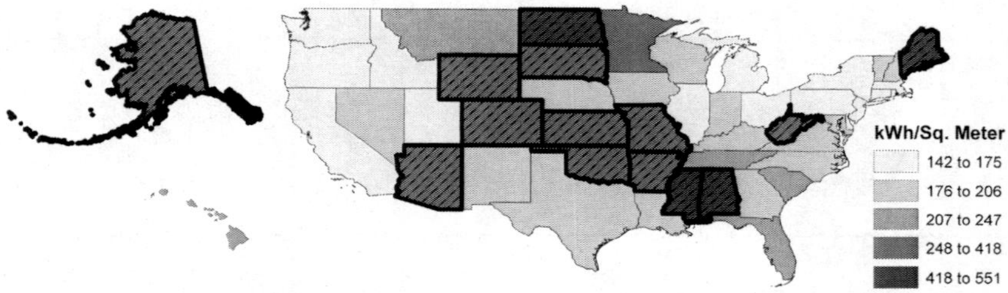

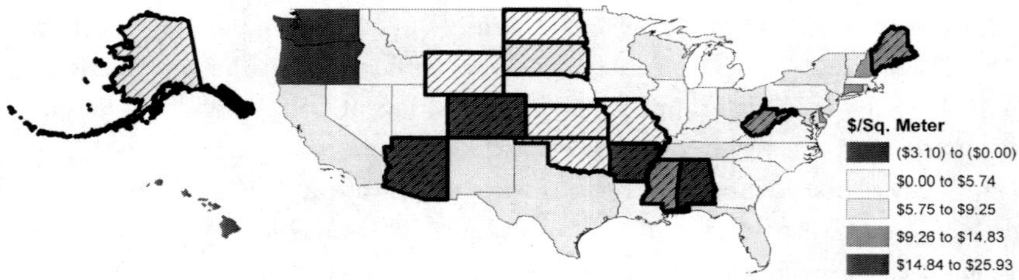

the net savings in energy use and life-cycle costs per unit of floor area. States with no code or ASHRAE 90.1-2001 [2] as its energy code (states with hatch marks) realize greater net savings in energy use per unit of floor area constructed than states that have adopted ASHRAE 90.1-2004 [3] or ASHRAE 90.1-2007 [4]. Net savings in life-cycle costs per unit of floor area are not highly correlated with the state's energy code. Some states with no state code (Alabama, Arizona, Arkansas, Colorado, Maine, Mississippi, and West Virginia) realize the greatest net savings per unit of floor area out of all states while others (Oklahoma, Kansas, Montana, South Dakota, and Wyoming) realize some of the smallest net savings.

The net savings estimate for the entire U.S. is the sum of net savings for each of the fifty states. **Table 6** shows the total net savings in energy use and life-cycle costs for one year's worth of construction for a 10-year study period aggregated at the Census region and national level. The greater the amount of new floor area constructed in a Census region, the greater the total net savings. The nationwide reductions total 34 441 GWh in energy consumption and $1.0 billion in life-cycle costs for a 10-year study period.

Discussion

This paper uses ASTM building economic standards (ASTM E917-05(2010) and E1074-09) to estimate the impacts that the adoption of more stringent energy codes for commercial buildings would have on building life-cycle costs. The results of

TABLE 6 Total reductions by census region from adoption of the LEC design, 10-year.

Census Region	Average Annual New Floor Area 1000 m^2 (1000 ft^2)	Energy Use (GWh)	LCC ($million)
Northeast	22 395 (241 056)	3437	125
Midwest	29 510 (317 641)	6621	136
South	67 299 (724 394)	16 427	488
West	36 144 (389 056)	7956	300
Total	155 348 (1 672 147)	34 441	1048

this paper are based on analysis of the BIRDS database. For this study, the performance of buildings designed to meet current state energy codes is compared to their performance when meeting a "Low Energy Case" (LEC) building design based on ASHRAE 189.1-2009 [7], which increases energy efficiency beyond the ASHRAE 90.1-2007 [4] design, to determine whether more stringent energy standard editions are life–cycle cost–effective in reducing energy consumption.

The analysis first estimates the net savings in energy use and life–cycle costs for a single building (3–story office building) in a single city (Knoxville, TN) to detail the approach used in the analysis. Constructing the 3–story office building to meet the LEC design requirements leads to annual net savings in energy use of 51 383 kWh and net savings in life–cycle costs of $41 350. The approach is then applied to five cities in Tennessee to estimate the net savings in energy use and life–cycle costs per unit of floor area constructed for each building category which has a representative building prototype. The average of the net savings per unit of floor area across the five cities for each building type is multiplied by the associated building category average annual construction from 2003 to 2007 to estimate the net savings in annual energy use (47.2×10^6 kWh) and life–cycle costs over the 10-year study period ($\$16.4 \times 10^6$) for construction represented by the prototypes. The building categories represented by the prototypes account for 58.4 % of all new construction in Tennessee. Assuming that the savings realized by the prototypes is representative of the entire new building stock, the aggregate statewide net savings in annual energy use (80.8 GWh) and life–cycle costs over the 10–year study period ($\$28.1 \times 10^6$). The approach is replicated for all 50 states and then totaled to estimate the national impacts from the nationwide adoption of the LEC design for one year's worth of construction for a 10–year study period: 34 441 GWh and $1.0 billion. Given the assumptions required to calculate the results, it is best to consider these calculations to be the general magnitude impacts instead of precise estimates.

Limitations and Future Research

The analysis in this study is limited in scope and would be strengthened by including sensitivity analysis and expanding the database. Sensitivity analysis is needed for at least two assumptions in the analysis. First, consider the assumed discount rate. Although 3 % is a reasonable discount rate, in real terms, for federal government investment decisions, it may be too low of a value for an expected real return on an alternative investment in the private sector. Sensitivity analysis on the assumed discount rate is needed to determine the robustness of the net savings results. Second, the current analysis assumes that the cooling load is met by equipment running on electricity while heating loads are met with equipment running on natural gas, which is not the typical fuel mix for some areas of the nation. The analysis should be expanded to include alternative fuel source options, such as heating oil use in the New England area or electricity in the South.

Additional data are needed to refine and expand the analysis. First, the study uses simple averages of energy use and life–cycle cost changes across all locations in a state. However, the amount of total floor area constructed will vary significantly from city to city. Future research could develop a weighted average of savings in a state based on the fraction of new construction by city. Second, the 11 prototypical buildings analyzed in this study are likely not representative of the entire building stock for each building type. For example, all high-rise buildings are not 100 % glazed. For this reason, the results should be considered as general magnitudes instead of hard estimates. Future research should include additional prototypes, such as the DOE Benchmark Buildings [28], in the database. Additionally, since existing buildings account for nearly the entire building stock, prototypes for energy retrofits to buildings should be incorporated into the BIRDS database as well. The state average energy cost rates do not control for local variation in energy tariffs. By using utility-level energy cost data, the accuracy of the estimates in BIRDS could be improved.

ACKNOWLEDGMENTS

Disclaimers: The policy of the National Institute of Standards and Technology and of ASTM International is to use metric units in all of their published materials. Because this paper is intended for the U.S. construction industry that uses U.S. customary units, it is more practical and less confusing to include U.S. customary units as well as metric units. Measurement values in this paper are therefore stated in metric units first, followed by the corresponding values in U.S. customary units within parentheses.

Certain commercial entities, equipment, or materials may be identified in this document in order to describe an experimental procedure or concept adequately. Such identification is not intended to imply recommendation or endorsement by the National Institute of Standards and Technology, nor is it intended to imply that the entities, materials, or equipment are necessarily the best available for the purpose.

References

[1] ASHRAE 90.1-1999: Energy Standard for Buildings Except Low-Rise Residential Buildings, ASHRAE/IESNA Standard Project Committee 90.1, ASHRAE, Inc., Atlanta, GA, 1999.

[2] ASHRAE 90.1-2001: Energy Standard for Buildings Except Low-Rise Residential Buildings, ASHRAE/IESNA Standard Project Committee 90.1, ASHRAE, Inc., Atlanta, GA, 2001.

[3] ASHRAE 90.1-2004: Energy Standard for Buildings Except Low-Rise Residential Buildings, ASHRAE/IESNA Standard Project Committee 90.1, ASHRAE, Inc., Atlanta, GA, 2004.

[4] ASHRAE 90.1-2007: Energy Standard for Buildings Except Low-Rise Residential Buildings, ASHRAE/IESNA Standard Project Committee 90.1, ASHRAE, Inc., Atlanta, GA, 2007.

[5] ASTM E917-13: Measuring Life-Cycle Costs of Buildings and Building Systems, *Annual Book of ASTM Standards*, ASTM International, West Conshohocken, PA, 2013.

[6] ASTM E1074-09: Measuring Net Benefits and Net Savings for Investments in Buildings and Building Systems, *Annual Book of ASTM Standards*, ASTM International, West Conshohocken, PA, 2013.

[7] ANSI/ASHRAE/USGBC/IES 189.1-2009: Standard for the Design of High-Performance Green Buildings, ASHRAE, Inc. and U.S. Green Building Council, Atlanta, GA, 2009.

[8] Commercial Buildings Energy Consumption Survey, 2003, "Commercial Buildings Energy Consumption Survey (CBECS) Database," http://www.eia.doe.gov/emeu/cbecs/ (Last accessed July 2009).

[9] US Department of Energy, 2009, "EnergyPlus Example File Generator, Building Energy Simulation Web Interface for EnergyPlus," U.S. Dept. of Energy National Renewable Energy Laboratory, http://apps1.eere.energy.gov/buildings/energyplus/ (Last accessed February 2009).

[10] Kneifel, J., "Prototype Commercial Buildings for Energy and Sustainability Assessment: Whole Building Energy Simulation Design," NIST Technical Note 1716, NIST, Gaithersburg, MD, 2011.

[11] Kneifel, J., "Prototype Commercial Buildings for Energy and Sustainability Assessment: Design Specification, Life-Cycle Costing and Carbon Assessment," NIST Technical Note 1732, NIST, Gaithersburg, MD, 2012.

[12] Database of State Incentives for Renewables and Efficiency, 2010, "Rules, Regulations, and Policies for Energy Efficiency database, building energy codes," www.dsireusa.org/ (Last accessed Jan 2012).

[13] US Department of Energy, 2009, "Building Technologies Program, State Energy Codes at-a-Glance," http://www.energycodes.gov/states/maps/commercialStatus.stm (Last accessed Jan 2012).

[14] RS Means CostWorks, 2009, "RS Means CostWorks Databases," http://www.meanscostworks.com/ (Last accessed 2010).

[15] Towers, M., Dotz, R., and Romani, L., *The Whitestone Building Maintenance and Repair Cost Reference 2008–2009*, 13th Annual ed., Whitestone Research, Santa Barbara, CA, 2008.

[16] U.S. Energy Information Administration, 2010, "Electric Power Annual State Data Tables," http://www.eia.doe.gov/fuelelectric.html (Last accessed 2011).

[17] U.S. Energy Information Administration, 2010, "Natural Gas Navigator," http://www.eia.doe.gov/oil_gas/natural_gas/info_glance/natural_gas.html (Last accessed 2011).

[18] Winiarski, D., Shankle, S., Hail, J., Liu, B., and Walker, A., "The Business Case for Sustainable Design in Federal Facilities: Appendix B: Energy and Construction Cost Estimates," Pacific Northwest National Laboratory and National Renewable Energy Laboratory, Richmond, WA, 2003.

[19] Belzer, D. B., Cort, K. A., Winiarski, D. W., and Richman, E. E., "Analysis of Potential Benefits and Costs of Adopting a Commercial Building Energy Standard in South Dakota," PNNL-15101, Pacific Northwest National Laboratory, Richmond, WA, 2005.

[20] Halverson, M.A., Gowri, K., and Richman, E. E., "Analysis of Energy Savings Impacts of New Commercial Energy Codes for the Gulf Coast," PNNL-16282, Pacific Northwest National Laboratory, Richmond, WA, 2006.

[21] US Department of Energy, 2009, "Building Technologies Program, EnergyPlus Energy Simulation Software," http://apps1.eere.energy.gov/buildings/energyplus/ (Last accessed 2012).

[22] Kneifel, J., "Life-Cycle Carbon and Cost Analysis of Energy Efficiency Measures in New Commercial Buildings," *Energy Build.*, Vol. 42, No. 3, 2010, pp. 333–340.

[23] Kneifel, J., 2011, "Beyond the Code: Energy, Carbon, and Cost Savings Using Conventional Technologies," Energy Build., Vol. 43, 2011, pp. 951–959.

[24] Rushing, A. and Lippiatt, B., "Energy Price Indices and Discount Factors for Life-Cycle Cost Analysis—April 2008, May 2008," NISTIR 85-3273-23, NIST, Atlanta, GA, 2008.

[25] Fuller, S. and Petersen, S., "Life-Cycle Costing Manual for the Federal Energy Management Program," *NIST Handbook 135*, 1995 ed., U.S. Dept. of Commerce, Technology Administration and NIST, Atlanta GA, 1996.

[26] Jarnagin, R. and Bandyopadhyay, G., "Determination of Weighting Factors for the ASHRAE Commercial Building Prototypes for Standard 90.1-2010," PNNL-19116, Pacific Northwest National Laboratory, Richmond, WA, 2010.

[27] McGraw-Hill Dodge Construction, 2014, "New Commercial Building Construction Data," http://construction.com/dodge/dodge-market-research.asp (Last accessed 1 Oct 2012).

[28] National Renewable Energy Laboratory, "U.S. Dept. of Energy Commercial Reference Building Models of the National Building Stock," TP-5500-46861, National Renewable Energy Laboratory, Golden, CO, 2011.

STP 1586, 2014 / available online at www.astm.org / doi: 10.1520/STP158620140031

Daryl DeJean[1] and Lee Stevens[2]

How ASTM E06.81 Standards Promote a Meeting of the Minds

Reference

DeJean, Daryl and Stevens, Lee, "How ASTM E06.81 Standards Promote a Meeting of the Minds," *Ideas to Impact: How Building Economic Standards Keep You on Track*, STP 1586, Robert E. Chapman and Michael N. Goodkind, Eds., pp. 184–201, doi:10.1520/STP158620140031, ASTM International, West Conshohocken, PA 2014.[3]

ABSTRACT

This paper focuses on how to make standards more usable and more valuable to key stakeholders over all life-cycle phases and in all sectors of the built environment. There is a need within the marketplace for well-established financial and economic evaluation tools both for traditional applications and for emerging trends—sustainability, Building Information Modeling, and other innovations—as they are mainstreamed into the built environment. The suite of standards from ASTM International's Subcommittee E06.81 on Building Economics fills that need. Specifically, these standards provide a full spectrum of value-enabling solutions to (1) accelerate the adoption of new building technologies and processes; (2) assist the evaluation of new energy efficiency and sustainability techniques; and (3) enhance the resource productivity of labor, materials, and energy. Taken together, these solutions can result in significant increases in efficiency and cost-effectiveness in the built environment. The standards also provide a framework that facilitates communication among key stakeholders, increasing the potential for identifying and implementing mutually beneficial solutions to complex problems. A series of case studies involving existing and new buildings and industrial facilities are used to illustrate how ASTM E06.81 standards can be applied. The case studies also serve to illustrate a sequence of steps that, if pursued, deliver greater value to key stakeholders. These steps involve (1)

Manuscript received March 26, 2014; accepted for publication June 4, 2014; published online July 14, 2014.

[1]MBA, President, Executive, Emerging Technologies Associates, Inc., San Diego, CA 92166, USA, e-mail: daryldejean@gmail.com

[2]MBA, President, Kingport Corporation, Santa Barbara, CA 93150, USA, e-mail: kingportco@gmail.com

[3]ASTM Symposium on *Ideas to Impact: How Building Economic Standards Keep You on Track* on October 5, 2014 in New Orleans, LA.

clarifying the capabilities of the E06.81 standards and simplifying presentation materials on how to apply ASTM E06.81 standards and (2) reaching and informing a wider audience by engaging stakeholders outside of the engineering community.

Keywords

financial and economic analysis tools, built environment, risk assessment, risk management, climate change risk, resource productivity, cost-effectiveness, building investment evaluation, cost classifications, market-disrupting innovation

Introduction

The purpose of this paper is to address the growing need in the marketplace for knowledge about and availability of a building-centric economic and financial evaluation process. This process will facilitate decision making and strategic planning in a rapidly changing global economy. Furthermore, it will allow for the mainstreaming of financial solutions, commonly used by the financial industry, to professions within the built environment.

ASTM International stands at a crossroads for promoting the use of its building economics standards, an objective, multidisciplinary, and multipurpose economic/financial evaluation framework. Promoting the use of these standards will (1) lead to a significant economic transformation of the global built environment and (2) trigger a needed financial effect on its projects and programs. Waves of sustainable best practices, technologies, and energy sources cause market disruptions in the building stock. Examples are market-disrupting technologies such as a light bulb that lasts seven years as compared to one year and the advent of intelligent or smart thermostats that remember the user's pattern of use [1].

ASTM E06.81 building economics standards [2] apply to every sector of the construction industry and every type of project, from a simple retrofit to the new construction of an entire campus. Potential users of ASTM E06.81 building economics standards include real estate investors; developers; property managers; tenants; and all supporting manufacturers, services, and governing bodies. As innovations enter the marketplace with their own assertions of cost-effectiveness, owners are confused by the diversity of methods used to illustrate the cost-effectiveness of these solutions. ASTM E06.81 holds the key to dispelling this confusion in the marketplace and demystifying the use of building economics standards.

In this paper, we introduce various case studies from public relations outreach and educational programs to actual use of the standards. These case studies demonstrate the need for the standards to be disseminated across many disciplines and across all sectors of the construction industry.

There are organizations, corporations, and government agencies that already conduct similar economic analyses. For example, the U.S. Environmental Protection Agency (EPA) ENERGY STAR program in partnership with Building Owners Managers Association International (BOMA) created several financial evaluation tools [3]. These evaluation tools include the calculators Cash Flow Opportunity, Financial Value, and Building Upgrade [4]. The purpose of these calculators was to allow the membership of BOMA and ENERGY STAR partners to have a consistent methodology for evaluating the financial effect of energy efficiency projects.

The opportunity exists for ASTM Subcommittee E06.81 to engage additional built environment stakeholders. The recommended approach is to demonstrate how the building economics standards, which are customized to the built environment, align with the financial tools used predominantly by investors and the commercial real estate community. This demonstrated alignment will allow for joint educational outreach to promote the importance of ASTM E06.81 standards in applying financial evaluation in the built environment.

We recommend a two-step approach to engaging these additional stakeholders and joining them in educational outreaches. The delivery of ASTM E06.81 standards must be simplified and focus on the application of the standards, in the simplest project as well as more complex new construction and renovation projects, in order to increase the appeal and success of the outreach effort within and outside of ASTM International.

The standards must be shown to align with accepted financial tools used in commercial real estate, financial services, and regulatory agencies to determine the cost-effectiveness of large-scale programs.

What Are ASTM Standards on Building Economics?

The ASTM E06.81 standards on building economics are a suite of 26 standards [2]. The standards are used by building owners and owner representatives to determine the cost-effectiveness of buildings, building systems, and building products. They apply to all phases of a facility's life cycle for both new construction and renovation, namely, planning, design, construction, operations, maintenance, and disposal.

ASTM E06.81 standards are a treasure trove. They are the most comprehensive set of guidelines, financial methods, and processes available in the built environment marketplace. They allow the standard user to perform the following:

1. Collection, presentation, and analysis of cost data
2. Determination of the value to the owner of alternative buildings and building systems across building disciplines
3. Assessment of life-cycle costs of buildings and building systems that are now of crucial importance given the growth of sustainable practices throughout the life of buildings
4. Selection of the economic methods or financial solutions that best fit the project at hand
5. Risk analysis and techniques for treating uncertainty that define different types of risks in the economic evaluation of a building or building system and its financial effects on the owner

The ASTM E06.81 building economics standards contain the same financial tools taught in basic college business accounting and financial management courses [5], whether at the undergraduate or the graduate level. The tools are used to evaluate capital expenditures as well as operations and maintenance expenditures. The standards draw upon the time-tested nature of these accepted financial tools. The main difference is that ASTM E06.81 standards are adapted to evaluate projects involving buildings and building systems.

Labeled as "Building Economics," ASTM E06.81 standards address inherent economic value by taking a rigorous approach to determining the cost-effectiveness of expenditures in the built environment. The title of "Building Economics" conveys the perception of the standards as tools focused on the microeconomic analysis of buildings and building systems.

Is There a Demand for Standards for Evaluating the Economic Value of Buildings and Building Systems?

There is a transformation about to occur in the built environment. As a result, we are at an inflection point in terms of how to perform a proper evaluation of built environment projects. This inflection point involves the traditional approach to a less competitive or less profitable decision or a newer, more comprehensive approach based on more critical thinking. By "critical thinking," the authors mean that the users can properly think through the situation, exploring all possible alternatives and associated costs, benefits, and risks. They must be able to communicate all of those alternatives to peers and decision makers to allow for consensus building within their organization. This is what ASTM E06.81 building economics standards enable.

Companies establish standards to ensure a level of quality and acceptable business practices for all their teams. Company standards are used in evaluating performance, managing risks, and improving operational efficiency. The DIN report "Economic Benefits of Standardization" found that industry-wide standards provide benefits for individual businesses by directly influencing a company's strategy, competitive advantage, and identification of key partnerships [6].

Today, there are numerous market forces demanding a more sustainable built environment. In analyses of the built environment, a whole-building and whole-system approach to designing a new building or managing an existing building, including managing the energy, water, and material resources, is recommended by many leaders in the built environment. The whole-building approach entails designing and integrating all resources, whether energy, water, material, or labor, in order to maximize the productivity of all resources used in a building project. This concept was introduced by Professor Ernst Ulrich von Weizsäcker and colleagues in the book *Factor Four: Doubling Wealth, Halving Resource Use* [7].

To properly achieve the integration of all resources, the analysis must cut across many disciplines, professions, and trades. This interdisciplinary aspect of the built environment has created a need. The need is for well-established standards for the financial and economic evaluation tools suitable for both traditional and emerging trends and applications.

Case studies from all over the world demonstrate how "resource productivity" can multiply four- or five-fold with the planned uses of interdependent resources within a building. These case studies are shared in the books *Factor Four: Doubling Wealth, Halving Resource Use* [7] and *Factor Five: Transforming the Global Economy Through 80 % Improvements in Resource Productivity* [8].

Strong Appeal for Global Standards in the Next Global Wave of Innovation

Companies value industry-wide standards that are established under "a formal, coordinated, consensus-based and open process" [9]. A consensus-building process is one allowing all group members to have the same level of influence. The DIN report "Economic Benefits of Standardization" found that companies believed if a majority vote were used in establishing standards, the standards would reflect the desires of more influential group members instead of those shared by the entire group [6]. Based upon the desire to embrace standards established by consensus, how do we convey the benefits of such standards so they become widely utilized and valued by key stakeholders?

ASTM International staff and committee member volunteers work hard to establish industry-wide standards that prove invaluable to all stakeholders involved in the production of research and the development of new technologies, products, and services, including those used in the built environment. ASTM International Committee E06 on Performance of Buildings offers a vast array of industry-wide standards allowing users to make better informed decisions when selecting building products, systems, and processes. The benefits of voluntarily embracing these standards can become incorporated in corporate strategies and operating procedures [10].

Industry-wide standards provide a level playing field for all manufacturers, producers, and services providers in a competitive market. Companies know that by meeting levels of performance specified by standards set by an impartial, unbiased third-party organization, they gain a competitive advantage. "Industry-wide standards form a collection of harmonized technical rules. This 'coding' of knowledge can help businesses cooperate and create strategic alliances. They act as an intermediary" [6].

According to the Business Standards Encyclopedia [11], other benefits of standardization include the following:

(1) Differentiating amongst suppliers to achieve more competitive pricing
(2) Confidence in the quality and reliability of solutions

(3) Increased interdepartmental communications to improve resource productivity
(4) Sharing best practices to improve design
(5) Establishing benchmarks for performance, quality, and safety
(6) Non-proprietary products allowing for interchangeability or working in other manufacturer products
(7) Reducing risks
(8) Reducing costs

The DIN report "Economic Benefits of Standards" found that industry-wide standards have a positive effect on the economy as a whole [6]. Industry-wide standards lead to two important questions: what type of detailed analysis should be performed to better understand the financial impact for projects in the built environment, and what role does standardization play in assessing the financial benefits of project development?

According to the British Standards Institution [12], "[b]usinesses today simply cannot afford to take an improvised, reactive approach to risk. Businesses rely on standards to identify and minimize risks." Industry-wide standards provide a means for effective analysis of investment choices, project prioritization, and management of the risks of capital expenditures. ASTM E06.81 standards create an unbiased methodology for the financial analysis of projects by eliminating reliance upon the financial justification provided by a specific manufacturer.

"The key lies in integrating standardization with corporate strategy or business strategy in accordance with the respective conditions for profit-making, while taking into account the nature and impact of standardization" [13].

There is a pent-up demand for readily available, simple, and easy-to-understand economic and financial tools for evaluating buildings and building systems for the owners of buildings and tenants or other stakeholders within the building industry. The demand is caused by the explosion of new building technologies being introduced in the marketplace as the "greening" of the built environment escalates [14]. Such economic or financial tools are readily available in the marketplace. However, aside from ASTM E06.81 standards, not one published source has formulated this important aspect of the decision-making process in a logical and easy-to-use manner. For example, the Internet provides no common thread or logic for internal rates of return, payback period, net present value analysis, and benefit-cost analysis. Furthermore, distinct disconnects exist among those who control the capital expenditures, those identifying projects, and the service and product providers seeking to promote new and emerging technologies—whether systems or products—for existing and new building stock.

It is interesting to note how the built environment suffers from an oversimplified economic and financial evaluation process to determine project worthiness. There is heavy reliance in certain sectors of the built environment on the financial method referred to as the payback method.

The payback method is used to determine "if a project recovers its investment cost. within its service life or a within a specified maximum acceptable payback period" according to ASTM E1121-12 [15]. The payback method commonly used in the built environment marketplace uses a zero discount rate and is referred to as "simple payback." ASTM E1121-12 warns the user of the standard of the limitation of the payback method, as "it ignores benefits and costs over the remaining service life of the project beyond the payback year" [15]. In spite of this shortcoming, simple payback continues to be used by facility managers and product and service providers in the built environment as a screening tool for investment decisions.

The overuse of simple payback has led to confusion in the marketplace and the slow adoption of advanced technologies for the following reasons:

1. It fails to take into account the economic and financial benefits that last well beyond the payback year.
2. The time value of money is not accounted for by not considering discounting to present value of any economic and financial benefits and costs beyond the first year.

There continues to be emphasis and reliance upon simple payback despite its shortcomings.

Why is there so much reliance on simple payback or other methods that lack consideration for other financial values? Based on observations and comments made by many stakeholders in the built environment, the authors are of the opinion that the payback method is used as a screening tool to determine whether the initial investment, commonly known as "first cost," is worthwhile.

The use of the simple payback method as a screening tool has extended to building owners, managers, and their service and product providers. These built environment stakeholders view buildings and building systems as cost centers and continuing sources of expenditures known as operations and maintenance. This mindset is changing with proper ongoing operations and maintenance becoming a sustainable building best practice.

There is well-documented research shared by ENERGY STAR partners in the public domain. This research demonstrates that managing operations and maintenance and capital spending on buildings, building systems, and processes can generate economic and financial benefits. These benefits include increased economic and asset value of a building and possibly extended life of a building [16].

In other industries, budgets are important. However, budgets are developed so as to ensure that all of the benefits of a project are considered. For example, in the shipping industry, a ship is built or refurbished with a stated purpose in mind. The purpose drives the design and the financial evaluation. Oftentimes there may be no hard monetary figure for an aspect or feature of a vessel. Yet, in the financial evaluation the fact that a need was met is valued. Additionally, the vessel as a capital project opens the horizon of the evaluation period to a longer time frame, resulting in the inclusion of many well-defined and targeted benefits over the vessel life.

A cargo vessel is designed and built to transport cargo in the most expeditious, cost-effective manner. There are certain features designed into the vessel that provide over its average 30-year life the opportunity for the owner to reap greater returns in the later years. Some of those features' values lie in the perception of the market that they may result in more business. Although these features are riskier and more difficult to assess, it is the evaluation method that provides the ability to model the long-term effects of their incorporation balanced against their higher first cost.

In the government sector, non–built environment projects are evaluated for mission readiness. Such projects are evaluated over the useful life of the project, considering the possibility of repurposing the project. The contribution to mission readiness is a very important consideration. Trying to meet a first cost analysis would result in potential elimination of the project, negatively affecting the mission of the government agency.

To date the built environment suffers from the first cost syndrome that stifles innovative projects. This way of thinking of the built environment as a financial burden rather than a contributor to a company's profitability is limiting. Thus a mind shift must occur. Perhaps the major contributing factor is that companies view the building as a secondary expense, because it simply "houses" their equipment and personnel. The value is in the equipment, people, and processes.

Other industries employ more holistic project evaluations. For example, in the shipping industry, certain features are necessary to provide a healthy, productivity-enhancing environment. Typically, it is not easy to put a hard financial number in the analysis phase that considers the impact the built environment has on productivity.

The authors of this paper are of the belief that the ASTM E06.81 standards on building economics contain guidelines that can bridge the existing communication gap.

Wider Application of ASTM E06.81 Standards Needs to Occur Now

Additional applications can be disseminated across the stakeholders who own, manage, and occupy buildings and other structures. Simple building projects such as the purchase by a restaurant owner of a new venthood with more innovative control features might not warrant dissection and scrutiny with a full-fledged life-cycle cost analysis. However, the too-common method of evaluation, years to payback, does not capture the longevity of the stream of cash savings that the new equipment could deliver in the restaurant. Often times, the payback period is used as a benchmark for accepting or rejecting such a simple investment without fully looking at the other benefits of employee productivity, longer life, etc.

The New Wave of Innovation Is Upon Us

As we move into a brave new world of climate change, sustainability, and greening of Planet Earth, the following significant developments are coming into play:

1. The next "wave of innovation" in manufacturing is the "emerging wave of green technologies" (see Fig. 4 on p. 13 of Ref 17 depicting the "Waves of Innovation" graph).
    a. What is a wave of innovation? It is the boost in economic activity resulting from the major introduction of technological breakthroughs such as steam power and the railway. The theory that the wave of innovation links long-term economic cycles and business cycles with major technological innovations is attributed to the research of economists and scholars Nikolai Kondratiev [18] and Joseph Schumpeter [19].
    b. Applying the wave of innovation theory to the economic trends of the 21st century, the Natural Edge Project [17] predicts that the next wave of innovation is upon us. Large manufacturing companies are greening their supply chains in response to regulatory mandates such as CALGreen 2013, also known as the 2013 California Green Building Standards Code [20]. There is consumer demand for sustainability complementing the push from environmental nongovernmental organizations. A prime example of how this wave has already transformed the marketplace is a highly effective public and private collaborative initiative known as the Sustainability Consortium. The Consortium, in conjunction with the research and development capability of Duke University's Nicholas Institute, is building a massive worldwide repository of information on the supply chains of many products. This repository will include building products eventually [21].
2. Where are we in the wave of innovation in the built environment? Bombarded with new building technologies that are constantly changing, the marketplace is in the early stage of adoption and is seeking solutions in the following areas:
    a. Acceleration and mainstreaming of solutions to deal with climate change risks and remediation in the following fields:
        1. Sustainability,
        2. Energy efficiency, and
        3. Renewable energy.
    b. Ways to quantify effects on the bottom line using a systematic evaluation of new sustainable solution costs, benefits, and risks of potential outcomes.
    c. A systematic approach to assess new and emerging technologies with tools other than "simple payback."
    d. Decision making, which needs to be shared at many levels within large and medium entities in both private and public sectors. The current flat organizational structure is causing a crisis in decision making for projects such as lamp retrofits, which can sometimes take 3 years to be approved. Decision making is caught in collective

thinking and delays. This often leads to incorrect decisions with potentially prohibitively costly fixes.

e. Project or program teams must be on the same page in evaluating projects/programs regardless of areas of specialization.
f. Economic recovery is demanded of all government leaders with little if any consideration of the tools used to accomplish this miracle.
g. Building owners and property managers are asking for simpler ways to value a project in a building.

3. Have ASTM members and stakeholders in the built environment caught the wave? How can ASTM International help move the wave to a higher level of economic productivity or wealth? ASTM International has been on the leading edge of the wave that has reached the built environment described in item 2.
    a. ASTM E60 sustainability standards have just been published by ASTM International. The publication of ASTM E60 sustainability standards continues to gain traction with Green Meetings standards cutting across all sectors of an economy.
    b. ASTM Committee E50 has created a building energy performance tool that is used by building owners and regulatory agencies
    c. The formation of the Built Environment Advisory Committee in March 2013 [22] was recognition within ASTM of the crucial role of standards in driving market adoption of sustainable new products and sustainable best practices.
4. A global shift to a new mindset, a "meeting of the minds," is necessary in economic and financial evaluation of emerging building technologies.
    a. The mainstreaming of sustainability in the economy has led building product manufacturers to introduce more sustainable products that have longer lives and that are more durable. Built-in obsolescence is no longer acceptable to the new consumer. This is often associated with a quantum leap in performance and even some intelligence or smart control.
        i The ensuing higher cost of the new technologies cannot be justified with traditional evaluation methods primarily because of the longer stream of benefits and costs derived from the investment in the new technology. "The payback method accounts for all monetary values associated with an investment up to the time at which cumulative net benefits, discounted to present value, just pays off initial investments costs" [15]. The payback method is still used to determine the cost-effectiveness of advanced technologies. The simple payback method ignores the stream of cash flows that results from the expenditure beyond the payback period. It immediately dismisses any avoided expenditures such as maintenance and disposal. This illustrates the blurred line between maintenance expenses and capital spending. As a result, there continues to be incomplete financial evaluation due to a lack of valuing such solutions in the built environment.

ii Another dilemma faced by building owners is that certain advanced technologies are treated as an expense as opposed to a capital investment.

iii Operations and maintenance represent an integral part of the Leadership in Energy and Environmental Design (LEED) certification process. The operating cash flow savings of a LEED-certified building relative to a non-certified building could be analyzed using ASTM E06.81 standards. A proper financial evaluation considers the long-term value of reduced capital spending and lower operating expenses over the life of the building. As a result, more investors would be convinced to look at LEED-certified buildings as more attractive investments.

The authors are of the opinion that ASTM E06.81 standards can be utilized to take into account all of the financial rewards of advanced technologies, providing a better understanding of the limitation of the simple payback method. ASTM E1185-12 provides the user with guidance on which method to utilize [23].

b. With the globalization of the marketplace, the need to assess business risk is greater. ASTM E1369-11 [24] and ASTM E2506-11 [25] provide a framework for incorporating business risk analysis in the built environment globally. ASTM E1765-11 [26] provides the methodology to help decision makers choose from among buildings and building systems in a way that optimizes trade-offs between cost-effectiveness and risks.

For example, climate change is gradually becoming accepted as a business risk. Assessment of the risk of climate change is starting to be incorporated in business school curricula. Geoffrey Heal, Professor of Social Enterprise in the Finance and Economic Division at Columbia Business School, is teaching our future business leaders and policy makers how to make decisions when faced with climate change risks "when we might know that certain things are more likely than others, but we have difficulty assigning a precise likehood rating to them" [27]; in spite of "the lack of quantitative information" that typically is available when assessing other business risks, Heal recommends four vital steps, "all of which are easily implemented, will remedy gaps in our economic system and remedy what economics text call market failures. They would improve the efficiency of our economic system, solve most of our environmental problems, revolutionize the relationship between humanity and the natural world, and give the planet a new lease on life" [28]. Heal strongly recommended in his keynote address at the Rio +20 Earth Summit that business leaders and policy-makers implement changes in the global economic system to improve the utilization of all resources, including our natural resources.

c. Beyond the scientific and academic communities, Wall Street is demanding transparency no matter what the government and business leadership believe in or how climate change risks are managed. The Dow Jones Sustainability Indexes and the Carbon Disclosure Project [29] publish disclosures of the sustainability initiatives of the largest publicly traded corporations. Included are the steps they have taken in managing outcomes associated with climate change risks, whether in the utilization of resources throughout the supply chain, mitigation of legal exposure, or product innovations that provide long-lasting solutions. The ASTM E06.81 standard compilation allows the user to properly collect, analyze, and present data in readiness for total disclosure of building performance.

A Case Study of a 2011 Town Revitalization Project

A town undertook a streetscape-beautification project. The purpose was to improve the overall appearance of Main Street. The project involved rebuilding the sidewalks, adding planters, and the installation of new light-emitting diode technology in place of the existing streetlights. One of the requirements was for Main Street to be reconstructed to be consistent with a historical New England main street. The town considered the total impact of the project in the revitalization of the appeal of Main Street. In evaluating the project, the town council looked beyond the paybacks of the improvements exceeding 3 years by considering the key benefit, the economic redevelopment of the downtown area.

As a result, businesses became engaged, completing upgrades to their storefronts. According to the business owners of shops on Main Street, the project had a positive impact by increasing tourist traffic in the area, bringing more revenue to the town and its small businesses, whose owners and employees for the most part lived in the town. In fact, residents and visitors commented on how pleasant and beautiful the Main Street area was and that it had led to the beautification of all of the shops by the merchants who benefited from the revitalization project. The town engineer did an outstanding job of presenting all of the positive financial and economic impacts of the total project instead of focusing on the energy savings of the streetlights, which would have had a payback in excess of 7 years.

Who Are the Stakeholders in the Built Environment Needing Economic Evaluation Tools?

There are many stakeholders interacting on new construction and retrofit projects and programs. The following key stakeholders would benefit from being made aware of the E06.81 building economics standards: building owners; business leaders and decision makers; property and facility managers representing ownership; developers and real estate investment companies; and sustainability teams or similar positions. The following could also be included in this list:

1. Financial services and investment bankers and insurance companies specializing in real estate
2. Engineering professionals (ASTM has reached out to many engineering associations and counts members among many engineering associations)
3. Green built environment and net zero building sponsors and stakeholders, a rapidly growing market
4. Trades that provide services in existing and new buildings: electrical, heating, ventilation air conditioning and cooling, plumbing, roofing contractors
5. Utility industry associations and representatives
6. Manufacturers of building products

Many of the above-mentioned participants in the built environment have expressed a need for better financial evaluation processes and tools. Finally, the decision-making process for new green built projects is spreading across organizations, business and government agencies, marketing, human resources, legal teams, and compliance teams. These disciplines are now becoming more ingrained in a complex decision-making process. In order for building projects to advance in the order of priority of capital spending, the cost-effectiveness of projects must be properly vetted using a validated and "standard" evaluation method. What better way to obtain validation than to refer to the E06.81 standards that are customized for use in the built environment!

Case Study of the Applicability of ASTM E06.81 Building Economics Standards in the Marketplace

As a part of updates to a program, E06.81 standards were presented in a statewide energy-efficiency training program to build awareness of these standards. The authors have served as trainers for the program throughout the state of California since 2008. We teach a business development seminar that includes a project development and financial evaluation section. The introduction of ASTM E06.81 standards was a natural progression in our updates to our curriculum.

The attendees tended to focus on the payback period. Every discussion they had with anyone emphasized the payback period. The payback method called "simple payback" is the primary financial tool used in the new construction and retrofit markets when considering the solutions selected. The payback method is used exclusively by utility companies to set rebates and incentives and to provide financing. At the same time, training seminars within the industry reinforce the use of this method because it is very simple to use. It is seen as a good screening tool for multiple projects and puts everyone on a level playing field in an industry where exposure to financial education is limited. In addition, the payback period was quoted by many contracting trades, engineers, and utility executives as the reason that energy efficiency projects were rejected. One of the authors of this paper managed over 200 facilities for a large manufacturer in North America, and that allowed us to deduce that owners and their representatives were using the payback period as a way to constrain projects that received funding.

The use of the payback period was particularly limiting for our audience, who factored into the calculations only reductions in energy bills to justify projects. There was no consideration of other financial and non-financial benefits that the owner and occupants might incur after project completion. Any projects with a payback of 2 years or more were automatically rejected, leaving many system solutions a challenge to justify in terms of cost.

Two years prior to introducing ASTM E06.81 standards, we introduced the U.S. EPA ENERGY STAR for Buildings and Plants Financial Resources [4]. The resources include three Excel-based calculators; by inputting data into the calculators, a user can generate, in less than 10 min, all of the values for payback period, return on investment, internal rate of return, and net present value. This invaluable part of the ENERGY STAR program is relatively unknown and not commonly used. An attractive feature of the ENERGY STAR calculators is that they are free. The use of the calculators is demonstrated for free as well as in webinar recordings that are available 24/7 on demand.

The introduction of the U.S. EPA ENERGY STAR financial calculators was successful. We were able to convince our seminar attendees that the partnership of large commercial building owners and a federal agency was a credible and reliable source. However, there was some remaining skepticism about the use of other financial ratios and methods besides the payback period.

In 2012, we began introducing ASTM E06.81 building economics standards into our curriculum. Many attendees commented that they had never heard of ASTM E06.81 standards and found them very useful. ASTM E06.81 standards helped validate the approach to evaluating energy efficiency projects. Participants found the "Section 2 Guides" [2] most applicable to their situation. Most connected with the following:

- ASTM E1185-12: Guide for Selecting Economic Methods for Evaluating Investments in Buildings and Building Systems [23],
- ASTM E1369-11: Guide for Selecting Techniques for Treating Uncertainty and Risk in the Economic Evaluation of Buildings and Building Systems [24], and
- ASTM E2204-11a: Guide for Summarizing the Economic Impacts of Building-Related Projects [30].

These three standards helped

- explain the advantages and disadvantages of each economic (financial) method,
- guide the user of the standards on when and where to use a particular method,

- provide a guide on how to model uncertainty and account for risks, and
- show how the different evaluation methods can be presented in a single document to facilitate the capital spending process internally.

As of January 2014, attendees had become very receptive to ASTM E06.81 standards and were excited to learn that they existed in a single document [2]. From San Francisco to San Diego, we continue to receive positive feedback that ASTM E06.81 standards combined with the ENERGY STAR financial resources are excellent resources for on-the-job proposal preparation. Most important, they helped counter objection and skepticism from facility management, construction team members, and owners who were not aware of ways of looking at building projects other than simple payback. The ASTM E06.81 standard that seemed to hit home most was ASTM E1185-12 [23].

How to Promote ASTM E06.81 Standards

To date, ASTM International outreach to the engineering community has been very successful with regard to technical standards. However, even within ASTM membership, there appears to be a very limited knowledge and awareness of ASTM E06.81 standards. An outreach effort must begin within ASTM International membership and then expand to other professions, trades, and governing bodies to influence building investments globally.

Based on our experience, we recommend a two-step approach to ASTM International leadership and interested stakeholders to promote the use of ASTM E06.81 building economics standards.

1. Simplify the presentation of ASTM E06.81 standards in outreach materials focusing on the guides.
   - Keep all material as free of mathematical formulas as possible.
   - Promote the guides separately from the other standards.
     - ✓ The standards that were of the most interest to our audience in the case study described above were the high-level guides. They communicated the tremendous value of the impartial and comprehensive ASTM E06.81 standards. The user of the guides learns how to match ASTM E06.81 standards to an organization's financial goals.
     - ✓ Develop a flyer for non-financial users of the guides with a brief description of the key features for each of the five guides.
2. Undertake public relations and marketing outreach to stakeholders in the built environment.
   a. Address organizations such as the International Facility Management Association (IFMA), BOMA International, the Institute of Real Estate Management (IREM), the American Society of Heating, Refrigeration and Air-Conditioning Engineers, the American Council for an Energy-Efficient Economy, the U.S. Green Building Council, and other non-profit organizations that have in place educational programs for their members.

b. Reach out to professionals, including regulators, involved in the development of codes and standards.
c. Other organizations that have a vested interest in building management are
   the American Institute of Architects
   Building Owners Management Institute International (http://www.bomi.org)
   CoreNet Global
   Counselors of Real Estate
   IREM
   IFMA
   the Society of Industrial and Office Realtors

Conclusion

ASTM International has a unique opportunity stemming from the dramatic changes occurring in the built environment. ASTM's opportunity is to promote ASTM E06.81 standard financial and economic evaluation methodology as a means for measuring cost-effectiveness. This methodology applies both to new and existing buildings and building systems and to new and emerging building technologies and processes. As result, ASTM will be viewed as the facilitator of the market transformation that leads to increased cost-effectiveness and resource productivity and as the creator of the evaluation standards that impact favorably the financial health of companies both domestically and globally.

ASTM E06.81 building economics standards both encompass all of the known financial methods and are customized for applications to the built environment. Embracing and promoting ASTM E06.81 standards allows for a natural progression by engaging a more diverse user group for these standards. The two-step recommendation for an outreach effort allows many building stakeholders to become aware of ASTM and the value of its standards across industries throughout the world.

References

[1] Gray, K., "4 Smart Thermostats That Save Money and Energy," *USA Today*, May 7, 2012, http://usatoday30.usatoday.com/tech/products/story/2012-05-06/smart-thermostats/54750056/1 (Last accessed 11 May 2014).

[2] ASTM Subcommittee E06.81 on Building Economics, *ASTM Standards on Building Economics*, 7th ed., ASTM International, West Conshohocken, PA, 2012.

[3] BOMA Energy Efficiency Program, "BOMA: Building Owners and Managers Association," https://www.boma.org/education/online-learning/beep/Pages/Curriculum%20and%20Registration.aspx (Last accessed March 23, 2014).

[4] U.S. Environmental Protection Agency ENERGY STAR Program, Buildings & Plants, "Financial Resources," http://www.energystar.gov/buildings/tools-and-resources/financial-resources (Last accessed 11 May 2014).

[5] Wild, J. J. and Chiappetta, B., "Capital Budgeting and Investment Analysis," *Financial and Managerial Accounting*, 2nd ed., McGraw-Hill, New York, 2007, pp. 948–960.

[6] DIN German Institute for Standardization, *Economic Benefits of Standardization, Final Report and Practical Examples*, Beuth Verlag, Berlin, Germany 2000.

[7] von Weizsäcker, E., Lovins, A. B., and Lovins, L. H., *Factor Four: Doubling Wealth, Halving Resource Use*, Earthscan, London, 1998.

[8] von Weizsäcker, E., Smith, M. H., Desha, C., and Stasinopoulos, P., *Factor Five: Transforming the Global Economy through 80 % Improvements in Resource Productivity*, The Natural Edge Project, Earthscan, London, 2009.

[9] ANSI News and Publications, "Media Tips and Case Studies Standards Overview: Avoiding Surprises—Some Thoughts on Standards," http://www.ansi.org/news_publications/media_tips/standards_overview_cont.aspx?menuid=7 (Last accessed 11 May 2014).

[10] Pauley, J., "Do Businesses Really Need Standards?" *Standardization News*, Nov/Dec, 2013, pp. 34–37.

[11] Business Standards Encyclopedia, "The Benefits of Standards," http://www.standards.bz/benefits.html (Last accessed 23 March 2014).

[12] British Standards Institution Group, "The Benefits of Using Standards," http://www.bsigroup.com/en-US/Standards/Benefits-of-using-standards (Last accessed 23 March 2014).

[13] Japanese Standards Association, "International Competition and Global Standards: Business Model for Standardization as Seen in Case Studies," compiled by the Study Group on Economic and Social Effects of Standardization, the Ministry of Economy, Trade and Industry, 2005, p. 3, http://www.iso.org/iso/home/standards/benefitsofstandards/benefits-detail.htm?emid=17 (Last accessed 23 March 2014).

[14] Clauson, D., "Greening the Built Environment," *ASTM Standardization News*, Nov/Dec, 2013, pp. 38–45.

[15] ASTM E1121-12: Practice for Measuring Payback for Investments in Buildings and Building Systems, *Annual Book of ASTM Standards*, ASTM International, West Conshohocken, PA, 2012.

[16] Office of Air and Radiation, U.S. Environmental Protection Agency, "Chapter 3: Investment Analysis," ENERGY STAR Building Upgrade Manual, 2008, http://www.wbdg.org/ccb/STAR/energystar_bldgupgrademanual.pdf (Last accessed 23 March 2014).

[17] Hargroves, K. and Smith, M., *The Natural Edge Project*, Earthscan, London, 2005, pp. 39–42.

[18] Kondratiev, N. D., *The Major Economic Cycles*, Richardson and Snyder, New York, 1984.

[19] Schumpeter, J. A., *Essays on Entrepreneurs, Innovations, Business Cycles, and the Evolution of Capitalism*, Addison-Wesley, Reading, MA, 1951.

[20] California Buildings Standards Commission, "2013 California Green Building Standards Code, CALGreen, California Code of Regulations Title 24, Part 11," 2013, http://www.ecodes.biz/ecodes_support/Free_Resources/2013California/13Green/13Green_main.html (Last accessed 11 May 2014).

[21] Duke University Center for Sustainability & Commerce at the Nicholas Institute for Environmental Policy Solutions, "Sustainable Products," http://center.sustainability.duke.edu/sustainable-products (Last accessed 11 May 2014).

[22] Enright, C., "Building Outside the Box," *ASTM Int. Standardization News*, Jan/Feb, 2014, pp. 28–31.

[23] ASTM E1185-12: Guide for Selecting Economic Methods for Evaluating Investments in Buildings and Building Systems, *Annual Book of ASTM Standards*, ASTM International, West Conshohocken, PA, 2012.

[24] ASTM E1369-11: Guide for Selecting Techniques for Treating Uncertainty and Risk in the Economic Evaluation of Buildings and Building Systems, *Annual Book of ASTM Standards*, ASTM International, West Conshohocken, PA, 2012.

[25] ASTM E2506-11: Guide for Developing a Cost-Effective Risk Mitigation Plan for New and Existing Constructed Facilities, *Annual Book of ASTM Standards*, ASTM International, West Conshohocken, PA, 2012.

[26] ASTM E1765-11: A Standard Practice for Applying Analytical Hierarchy Process (AHP) to Multiattribute Decision Analysis of Investments Related to Buildings and Building Systems, *Annual Book of ASTM Standards*, ASTM International, West Conshohocken, PA, 2012.

[27] Heal, G., "Uncertainty, Risk, and Climate Change," Columbia Ideas at Work Podcast, July 29, 2013, http://soundcloud.com/ideasatwork/Geoffrey-heal (Last accessed 11 May 2014).

[28] Heal, G., "Planetary Economics," Columbia Ideas at Work, June 26, 2012, http://www8.gsb.columbia.edu/ideas-at-work/publication/790/planetary-economics#.Ux3VPI6V8yg (Last accessed 11 May 2014).

[29] "Dow Jones Sustainability Indices in Collaboration with RobecoSAM," http://www.sustainability-indices.com/about-us/dow-jones-sustainability-indexes.jsp (Last accessed 11 May 2014).

[30] ASTM E2204-11a: Guide for Summarizing the Economic Impacts of Building-Related Projects, *Annual Book of ASTM Standards*, ASTM International, West Conshohocken, PA, 2013.